World Encyclopaedia of
AERO ENGINES

World Encyclopaedia of
AERO ENGINES

Bill Gunston

 Patrick Stephens, Wellingborough

First published in 1986
Reprinted August 1986
Reprinted February 1987

British Library Cataloguing in Publication Data

Gunston, Bill
    World encyclopaedia of aero engines.
    1. Airplanes—Motors—History
    I. Title
    629.134—35—09   TL701

    ISBN 0-85059-717-X

*Patrick Stephens Limited is part of the
Thorsons Publishing Group*

Printed and bound in Great Britain

Introduction

As my wife used to fly gliders I had better go easy on the notion that a flying machine is nothing without an engine, but the importance of engines is obvious enough. In recent years popular books about aircraft have seen almost explosive growth, but these books inevitably only skate over the surface. The authors of these works would no more think of describing the engine(s) than they would think of explaining how a wing gets its lift.

It was no small surprise when the British *Aeroplane Monthly* conducted a readership survey and found that the majority of responding readers had ticked the box labelled 'Aero Engines'. This triggered a short series of articles on some classic engines, which in turn led PSL to wonder if there might be a market for a book.

I am grateful to the publisher, in particular to Editor-in-Chief Bruce Quarrie and Graham Truscott. There have been very few attempts since 1921 to produce a book which sets out to give a quick guide to *all* significant aero engines. It is only just possible, in a book of affordable size. Obviously one has to be selective, and I have tended to include the companies and other producers whose engines were in some way important. I have attempted to include also the lesser-knowns from the big companies, and I would like to thank many people who have helped me track down scarce photographs. I may have to face criticism from people asking why I did not include particular favourites of theirs (apart from a mass of minor pre-1914 engines most omissions are in the low-power or home-built category), and I apologize in advance. You just cannot include them all.

This book is concerned with engines for aeroplanes and helicopters, and has little to say about missiles or spaceflight. It is logically arranged in alphabetical order of manufacturer, each entry then attempting to trace the history chronologically. I have concentrated on the engines, and often omitted mention of the applications. I suspect there are few people who wish to read about the Merlin or J79 who have no idea what aircraft these engines powered; in any case, it would

add many pages just to list them all. Most readers will already be familiar with the applications, and will, I hope, use this book to provide background detail.

Where possible I have mentioned key people involved and it will be obvious to the thinking reader that, behind the story of every engine, there is a wealth of drama, violent argument, moments of triumph and catastrophe, and often nagging doubts, or uphill struggles that seem to get nowhere. In a few cases I have commented on the way that men with highly-paid secure jobs threw them up because of their inability to penetrate the closed minds of a technically illiterate board of management. Even today one occasionally finds companies—not often in aviation—run by people who do not understand the firm's past success and have no idea how to sustain it. One day the Oxford Union might like to debate 'That this house considers that every manufacturing company should be run by (a) accountants or (b) engineers (please state your preference)'. Lord Hives of Rolls-Royce inclined to the engineers, saying, 'If the engineers are wrong, then we are all wrong'. Others might claim that too many engineers and too few accountants brought about the bankruptcy of that great company (though long after Hives' time). But one cannot even skim across stories of modern aero engines without appreciating that this is no place for the small man, nor for the fainthearted. To make even a seemingly trivial change to a major high-thrust turbofan swallows up not just a few million dollars or pounds but a few hundred million. A great engine designer, Sir Stanley Hooker, said of a famous banker, 'We added a zero to his stature; he used to think £5 million a lot of money, but after a few weeks on the RB.211 he came to understand that £50 million is peanuts'.

It is all very well merely to *read* about such sums. Spare a thought—I almost wrote 'spare a copper'—for the people who actually have to find the money. Afterwards everything seems obvious, and one wonders how company presidents or investors or ministers could be so stupid. At the time, without a fully certificated crystal ball, it is a bit harder. Let's imagine you are a representative of the Quebec government, or of a group of Montreal investment banks, and a chap named Thor Stevenson arrives, saying he wants to borrow about $10 million, to build an aero engine. It is to be a small turbine in the 600-hp class to replace the famous Pratt & Whitney Wasp designed in 1925. There were lots of Wasps once, but there are no small turbines to speak of. To the obvious question, 'How many of these engines have you sold?' Stevenson replies 'We know we can sell 40. Beyond that we are certain there is a big market'. What do we do with the $10 million? Most sensible people would reluctantly say it was too much of a gamble; and that is just what they did say, back in 1958. But somehow the first PT6 got built. Then one was sold, to power a Hiller helicopter that was promptly cancelled. Phew! What a good thing we didn't part with our $10 million.

I like this story, because, as I write, the total number of PT6 engines sold is about 23,900. At least, that is what it was a few days ago. They've probably sold a few dozen since I began to write. Like I said, aero engine development is a great place for people who can get it right.

Of course, if you write a book about the many hundreds of types of aero engines, you cannot in the same book explain how they work, or present any kind of coherent history of the species. Which was the first aero engine? One could argue indefinitely, but one thing I have always found puzzling is why the many great would-be aviators of the 19th century spent so much effort trying to create massive and complicated engines to turn a propeller or flap the wings. All they needed was a bit of thrust for a short time, and they could get that, reliably and at modest cost, with a few of Mr Congreve's rockets. Even the few inventors who did think of rockets then spoilt it by trying to make rockets fed by *a steam boiler*!

For the record, the first man to drive a piston down a cylinder by igniting petrol vapour—then a very rare commodity— was Robert Street in 1794. But he was a bit late. Three years earlier Samuel Barber had invented the gas turbine. Or maybe we should give the credit to Hero of Alexandria, around 100 AD?

I have used both Imperial and metric units in this book depending on which were published by the original manufacturer. I have avoided abbreviations except: hp, horsepower; lb, pounds of thrust; h and s, the SI abbreviations for hours and seconds; TBO, time between overhauls; sfc, specific fuel consumption; rpm, revolutions per minute; and pr, pressure ratio.

Bill Gunston
Haslemere, Surrey.
October 1985

Acknowledgements

The author would like to thank the following for providing photographs: at Allison, John A. Beetham; Avco Lycoming Williamsport, Ken Johnson; John Batchelor, whose assistance was exceptional; Fleet Air Arm Museum, who photographed their Fairey Prince; General Electric, Dwight Weber (and reviewers of the text); Roger Hargraves; Ray Holl CEng; Mike Hooks: Jane's Publishing Company, Anne Corfield; NASM, Pete Suthard, who found the installed IV-2220; Orenda, David Roberts; Pratt & Whitney, Harvey Lippincott and Robert E. Weiss (and text reviewers); Pilot Press, who found the Jendrassik; Rolls-Royce, Mike Evans, John Heaven, Douglas Valentine and Jack Titley (plus Dominic Leahy and many reviewers); Hugh Scanlan, former Editor of *Shell Aviation News*: The Science Museum; John Stroud; Shorts, Tom Goyer (who tracked down the Pobjoy picture); SNECMA, Philippe Dreux and Martine Messauer; Ann Tilbury; Teledyne Continental Motors, J.L. Lawhead; and Sir Frank Whittle who reviewed the Power Jets text.

A

ABC (UNITED KINGDOM)

ABC Motors, the successor to ABEC (All-British Engine Company), was founded in 1911 by Granville Bradshaw, one of the flamboyant extroverts common in the early years of aeroplanes. He was a better salesman than engineer, though he designed his engines himself. The first aero engines were a 40 hp water-cooled in-line and a 100 hp V-8. The latter failed to appear in time to power the Flanders B.2 at the 1912 Military Trials, and the 40 hp version had to be substituted. Despite this the B.2 flew at 55 mph with three adults aboard. In 1915 a small 5 hp flat twin drove the APU (auxiliary power unit) generator on the Pemberton-Billing (Supermarine) P.B.31 Night Hawk anti-Zeppelin patrol aircraft. A new range of engines was started with the Mosquito 6-cylinder air-cooled static radial of late 1916. This had cylinders of 4.5 in bore and 5.9 in stroke, machined from steel forgings and with overhead valves. It was designed for 120 hp but completely failed its bench tests. In its stead Bradshaw produced the 7-cylinder Wasp, with similar cylinders and designed throughout for ease of production. Weighing 290 lb dry, and of 657 cu in capacity, it ran reasonably well and gave 170 hp. It powered numerous prototypes, and was developed into the 200 hp Wasp II.

This encouraging background was unfortunate. When Bradshaw produced his bigger Dragonfly in 1917 he claimed it would replace every other engine. An enlarged Wasp, its 9 cylinders were of 5.5 in bore and 6.5 in stroke, giving capacity of 1,389 cu in. It was claimed to give 340 hp for a weight of 600 lb, and was clearly simple and easy to mass-produce. The director of aeronautical supplies, Sir William (later Lord) Weir, took the rash decision to standardize on the Dragonfly in almost all new fighters and bombers planned for 1918. Production on a colossal scale was organized, and soon 1,000 had been delivered and hundreds more were appearing each month. Only then did it emerge that the engine was a disaster. It gave only 295 hp, weighed 656 lb, quickly became red-hot (it had what S.D. Heron called 'probably the worst example of air cooling ever used on a production aircraft engine') and, as it had, by chance, been designed to run at the crankshaft's critical torsional vibration frequency, the unhappy result was that this component usually broke after an hour or two. So severe was the vibration that propeller hubs quickly charred from the friction. The British aircraft industry was thrown into chaos, the old Bentley BR.2

The author's long talks with Captain Norman Macmillan and Major Oliver Stewart MC left him in no doubt that the ABC Dragonfly would have necessitated a frantic re-engining programme for thousands of aircraft had World War 1 lasted into 1919. (Engine in London's Science Museum.)

and other engines were put back into production and Bradshaw was, to say the least, unpopular.

His one saving grace was his family of small opposed engines, which continued in December 1915 with the Gnat. Bore and stroke respectively were 4.3 in and 4.7 in, giving capacity of 139 cu in, and the engine was rated at 45 hp in its geared version. This failed its tests, so the P.V.7 (Grain Kitten) had to use the 35 hp direct-drive version. Some hundreds of these were used in almost all the British 'cruise missiles' of the day, as well as the manned Sopwith Sparrow and a few other types. From the Gnat came the Scorpion of 1921, with small cylinders of 3.6 in bore and stroke, giving 73 cu in capacity, and rated at some 24 hp. Many were used in ultra-lights of the day, the Scorpion II having bigger cylinders of 4 in bore and 4.8 in stroke and giving up to 40 hp. Four Scorpion II cylinders were used in the 82 hp Hornet of 1924. In World War 2 ABC produced APUs.

Aichi (JAPAN)
This aircraft firm built engines from 1927, the main wartime type being named Atsuta after the factory; it was a licensed DB 601. The Ha-70 was a twinned Atsuta 30 rated at a claimed 3,400 hp.

Alfa Romeo (ITALY)
The famous automotive company of Alfa Romeo bought a licence for the Bristol Jupiter in 1925, and via the Pegasus and Mercury developed the 125.RC series of 1933. This was a Pegasus with only two valves per cylinder,and thus the company was able to do what eluded Fedden at Bristol: double up and produce an 18-cylinder two-row version. This engine, the 135.RC34 Tornado of 1935, was eventually developed to 1,800 hp. The company also produced an original 9-cylinder radial in 1930, the 215 hp D2, developed to 240 hp by 1935. The DH Gipsy Six was the basis of the Alfa 115 family of 1936-56, but the 121 inverted V-8 had only 110 mm stroke, and was typically rated at 380 hp. In 1952 licence production of the DH Ghost turbojet started, followed by such engines as the GE J79, J85, T58, T64 and T700. Today the company is again developing its own engines, based on the simple AR.318 turboprop in the 600-hp class originally designed in partnership with Rolls-Royce. From this has been derived the ARTJ.140 turbojet, with thrust planned to extend up to 1,984 lb.

Allis-Chalmers (USA)
This famous engineering company was a licensee of Brown-Boveri of Switzerland producing industrial gas turbines. This brought it into the field of aircraft gas turbines. In October 1941 it was invited to work on a ducted fan (turbofan) for the US Navy, and

received a contract in January 1942. Progress was so slow that in June 1943 the Navy cancelled the contract and instead ordered licence-production of the DH Goblin turbojet as the J36. This was flown in the Lockheed XP-80, Curtiss XF15C-1 and Grumman XTB3F-1, but again the company failed to perform on schedule and the contract was cancelled after the delivery of seven engines.

Allison (USA)
This Indianapolis company got into aviation in 1926 with the V-1410, an inverted version of the wartime Liberty with air-cooled cylinders. It did extremely well on test in a DH-4. In 1927 came the 765 hp two-stroke diesel airship engine for the US Navy, and it was largely Navy efforts to find a US replacement for the Maybach in its airships that led to the contract of 28 June 1930 for a single example of the proposed V-1710-A, to deliver 650 hp at sea level, that launched the only US liquid-cooled production engine of World War 2.

The original V-1710 was a sound V-12 with each bank of 6 cylinders cast as one block of light alloy. Bore and stroke were 5.5 and 6 in, respectively, giving the 1,710 cu in displacement denoted by the designation. There were four valves per cylinder and an excellent reduction gear based on work going back to the company's copy of the RR Eagle gear in 1924. Allison was a pioneer of plain bearings with lead/copper backed by steel, and the V-1710 was also the first engine designed from the start to use not water for cooling but ethylene glycol, a viscous fluid pumped round under pressure at much higher temperature (typically 250-300°F, 121-149°C) and thus carrying away the excess heat through a smaller radiator giving reduced aircraft drag. The V-1710 was also planned from the outset to use a turbosupercharger, but this was not needed for airships. The engine first ran in August 1931 and was ready for delivery on 12 February 1935. On this very day the giant airship *Macon* was destroyed, the Navy cancelled and James Allison sold out to General Motors. Allison remains a GM division to this day.

Under new chief engineer R.M. Hazen, the engine was redesigned in detail, and the Army by this time began to think it might mature quicker than the Continental, so long regarded as the top future fighter engine. By March 1937 Hazen's C8 version passed a type-test at 1,000 hp on 87-octane fuel; this was more than any Merlin had done and the US engine was also lighter. The turning point was when Don Berlin, chief engineer of Curtiss Airplane Division, put a turbocharged C8 into a modified P-36. The resulting XP-37 reached 340 mph at 20,000 ft, a major jump in speed and height which confirmed Army belief in the liquid-cooled engine. Thus the Allison engineering

The final mass-produced wartime Allison was the V-1710-F30L, used in the P-38L with remote turbos (not included in the dry weight of 1,395 lb), which was rated at 1,475 hp for take-off. It came in F30L and F30R forms, the P-38 having handed (opposite-rotation) propellers.

staff of 25 found themselves suddenly not only clearing the V-1710 for production but also developing the pusher version for the Bell XFM-1, the shaft-drive model with remote reduction gear (with 37 mm cannon passing through it) for the Bell XP-39, the regular model for the Curtiss P-40 and left/right handed versions with remote turbos for the Lockheed XP-38!

In 1939 the P-40 was selected for major production, at last bringing a reward for nine years and $2 million invested equally by the Army/Navy and Allison. Throughout the war the V-1710 remained in volume production, some 47,000 being built. A few dozen

were of the advanced E-series, with various forms of two-stage supercharger to make up for the lack of a turbo (which was used only on P-38 engines). The V-1710-119 drove the lightweight XP-51J Mustang at 491 mph and after the war the 2,300 hp V-1710-143/145 replaced the Packard Merlin in the F-82 Twin Mustang. In general the British engine's superior supercharger(s) always kept it ahead at high altitudes, but the author can testify to the Allison's outstanding smoothness at low levels to which the P-39, P-40 and early P-51s were usually consigned.

Allison wisely refused the Army's request to build a new 2,000 hp engine in 1937; they could barely develop the V-1710. Instead it coupled two V-1710s together to form the V-3420, flown in a B-29. Thousands were ordered to power the Fisher XP-75 escort fighter; then this programme was cancelled. Piston-engine work was abandoned in 1947.

Gas turbines

In September 1945 responsibility for the General Electric J33 turbojet was transferred to Allison, which thereafter produced the engine for many fighters, attack aircraft and trainers, and also carried out increasingly sure development including afterburning versions. In direct descent from Whittle's W.2B, the J33 typically had a diameter of 49.3 in, dry weight of 1,786 lb or 2,465 lb with afterburner, mass flow of

Below left *Handed over from GE, the J33 was further developed by Allison and built in very large numbers. This is a J33-16A of 6,350 lb rating, engine of the F9F-7 Cougar.*

Below *The Allison V-3420 was essentially two V-1710s with the inner blocks set 30° apart. There were several versions, this V-3420-A15 having both crankshafts geared to a single propeller. Weighing only 2,600 lb, this engine was rated at 2,600 hp at 3,000 rpm, which on a 15-minute basis could be maintained to 25,000 ft.*

87 lb/s and maximum thrust of 4,600 lb or 5,400 lb with water injection or 7,000 lb with afterburner.

In September 1946 Allison was also handed the General Electric J35, and this was made in even greater quantity. All versions had an 11-stage compressor handling 85 lb/s at a pressure ratio of about 4.9, eight combustion chambers and a single-stage turbine. Diameter was 37 in, weight about 2,260 lb or 2,695 lb with afterburner, and maximum thrust 5,600 lb, or 7,500 lb with afterburner.

Allison had sufficient confidence in August 1948 to start total redesign of the J35, at first (to ease the release of funds) calling this the J35-A-23 but later giving it the new designation J71. This had a 16-stage compressor, mass flow of 160 lb/s at a pr of 8, cannular combustor with 10 cans and a 3-stage turbine. It powered various aircraft and missiles, including the Douglas B-66 and McDonnell F-3.

In 1944 Allison began the long struggle to produce a turboprop. With Navy funds the T38 (Model 501) was designed, with a 19-stage compressor (pr, 6.3), 8 combustion chambers and a 4-stage turbine. This ran in 1947 and at a power of 2,250 shp flew in a company B-17 in April 1949. At 2,763 hp it powered the Convair Turbo-Liner, an Allison-owned CV-240. Two T38 power sections were joined to one remote gearbox in the T40 (Model 500), which flew at a design rating of 5,500 shp in the Convair XP5Y flying boat on 18 April 1950. Later T40s were developed to 7,500 hp in the XFV-1 and XFY-1 VTOL fighters.

From the T38 Allison derived the T56, funded by the USAF for the C-130. With only 14 stages this started life at 32 lb/s at pr of 9.25, and despite running at a constant 13,820 rpm instead of the previous 14,300, gave 3,460 hp plus 726 lb thrust, or 3,750 ehp. In September 1954 the USAF ordered 288 T56-A-1 engines for the C-130A. Today the T56 is still in

Above *Looking much like the first T56 of well over 30 years previously, the T56-A-427 is coming into production at 5,250 shp, to power the E-2C Hawkeye. In the commercial Model 501 the propeller is below the axis of the engine, the inlet being above, and this carried over from the Electra into the T56-A-14 used in the P-3 Orion.*

Below *Not an engine, but a shape becoming so important it ought to be in the book: the Hamilton Standard 9 ft propfan which will fly on a Lockheed-modified Gulfstream II powered by an Allison 501-M78 (note modified mock-up inlet).*

volume production with 14,460 examples delivered, including commercial Model 501 engines for the Lockheed Electra (which has the air inlet above the spinner, as do the versions for the P-3 Orion). The current T56-A-15 has almost the same mass flow and pressure ratio as in 1954, but runs hotter and is rated at 4,591 shp (4,910 ehp), for a weight of 1,825 lb. By 1985 the T56-427 was in production for the E-2C Hawkeye rated at 5,250 shp and with 13 per cent lower sfc. A free-turbine derivative in the 6,000-shp class is the Model 501-M80C, designed for all-attitude operation in the V-22 (previously called JVX) tilt-rotor aircraft in 1987. The YT701 (501-M62) is a largely new modular free-turbine engine originally developed in 1970-75 for the XCH-62 HLH (heavy lift helicopter). After almost a decade of marking time this is again an active programme, and this increased-airflow engine will fly the XCH-62 in 1988 at a rating of 8,079 shp. The YT701 served as the structural basis for the 501-M80C, and another derived engine is the 501-M78 turboprop due for certification in 1986 as the PTA (propfan test assessment) engine at aircraft flight Mach numbers up to 0.8.

In June 1958 Allison received a US Army contract for development of a small but very advanced gas turbine in the 250-hp class, the T63. Unlike previous small turbines it had 6 axial stages of compression upstream of the centrifugal impeller that flung the air, at 3.1 lb/s at pr of 6.3, into two diffuser pipes leading to the rear of the single large reverse-flow combustion chamber with a single fuel nozzle. From the combustor the gas continued to travel forwards through the 2-stage gas-generator turbine and separate 2-stage power turbine before exiting through left and right exhaust stacks facing diagonally upwards. This engine created a great impression and appeared to have the world at its feet, but for more than 10 years virtually the whole output was tied up in the form of the T63-A-5 turboshaft for the Hughes OH-6A US Army helicopter and subsequently for the same customer's Bell OH-58A. The A-5 was rated at 250 shp, but by 1965 the A-5A raised this to 317 shp, and the same rating was agreed for the commercial Model 250-C18 and 250-B15 turboprop. In the latter the power section was inverted so that the jetpipes discharged diagonally downward.

Thanks largely to continued sales of the JetRanger and Hughes 500 the production total rose steadily, to 11,000 by 1976 and to almost 17,000 by 1985. This includes substantial numbers of the B17B and B17C turboprops, rated at 400 and 420 shp respectively, and the 420 shp C20 turboshaft. In December 1977 Allison certificated the 250-C28, with a totally new mechanical layout the main feature of which was elimination of the axial compressor; the single centrifugal stage actually achieves the increased pr of 8.4, besides raising airflow to 4.45 lb/s, giving a 30-minute rating of 500 shp. Many features improve reliability and reduce noise and emissions, including IR (heat)—important for combat helicopters. Today the C28 and plain-inlet C28B of 550 shp remain in production along with the 700 shp C30 with a further-

The C20R is the 1985 version of the original series of Allison 250 turboshaft engines, with 6 axial stages upstream of the centrifugal. It has a 5-minute rating of 450 shp.

developed compressor and 735 shp C34 which introduced a new single-stage gas-generator turbine. The complete redesign of the C28 was accompanied by a sharp reduction in turbine temperature which opened the way to further growth, and even the C34 still runs cooler than the early versions. The C28, 30 and 34 are, however, physically bigger engines, weighing roughly 100 lb more than the 150 lb of the first turboshaft versions.

In 1984 Allison's 4-year effort on the GMA (General Motors Allison) 500 had made this one of two finalists in a new US Army competition for an advanced engine in the 800-hp class, for the LHX family of helicopters and possibly even armoured vehicles. Features include 2-stage centrifugal compression, reverse-flow annular combustor and two-stage axial gas-generator and power turbines, the former with single-crystal air-cooled blades. This work has led to the Allison/Garrett proposal described below. Moreover Allison consider it too good not to take further themselves and in 1985 marketing began of the Model 280 commercial helicopter engine in the 800-shp class.

Allison/Garrett (USA)

So important is the US Army LHX helicopter programme that US engine companies are collaborating to try to win the propulsion contracts. Allison/Garrett LHTEC (Light Helicopter Turbine Engine Co) has been formed to propose the ATE 109, a 1,200-shp class engine which combines elements of Garrett's F109 and TSE109 with technology previously demonstrated by Allison. The rivals are GE and a joint effort by Avco Lycoming and P&W.

Alvis (UNITED KINGDOM)

Famed builder of quality cars, Alvis took the decision in August 1935 to take a licence for various Gnome-Rhône engines, notably the 14K, thus completing the circle begun by GR's purchase of a licence from Bristol in 1921! Alvis boldly conjured up Greek names—Pelides, Pelides Major, Alcides and Maeonides for an impressive family of Anglicized GR engines, most of them with 14 or 18 2-valve cylinders of 5.75 in stroke and 6.5 in bore, but they failed to penetrate the market despite the accelerating RAF expansion. The one engine that did make it was largely Alvis's own creation, led by Captain G. Smith-Clarke: the neat Leonides. This was a conventional small radial with 9 two-valve cylinders of 4.8 in bore and 4.41 in stroke, giving capacity of 718.6 cu in. The prototype weighed 693 lb and ran in December 1936 at 450 hp. In 1938 the firm arranged with Airspeed (1934) Ltd for their test pilot, George Errington, to carry out flight tests. These were done

With a diameter of only 41 in, the Alvis Leonides was a very attractive engine in the 500-600 hp class.

using the much-rebuilt Bristol Bulldog *K3183*, which had previously done the same for the Napier Rapier.

Leonides development was continued during the war at a reduced rate, and following testing in an Oxford and in Consul *VX587* (itself a converted Oxford) Alvis was ready in 1947 to market the engine as the Series 500 (502 and 503 and subtypes) for aeroplanes and Series 520 for helicopters. The first production applications were the Percival Prince, flown in July 1948, and the Westland Sikorsky WS-51 and Dragonfly helicopters with the 550 hp Leonides 521/1 and successor versions with direct drive, centrifugal clutch and fan cooling. All Leonides had an efficient low-pressure fuel-injection system, and a vast range of accessories and equipment suited them to many applications. This excellent engine was Britain's last high-power production piston aero-engine, the last being delivered in 1966. From 1959 a proportion of production was of the Series 530 (mainly 531 for Twin Pioneers) with stroke increased to 4.8 in, rated at 640 hp.

From 1951 Alvis developed the 14-cylinder 2-row Leonides Major, eventually certificating the Mk 702/1 for aeroplanes at 875 hp and the 751/1 for helicopters at 850 hp. The only significant production was of Mk 755/1 inclined-drive engines for Whirlwind Mks 5,6,7 and 8. Alvis spent much of the 1950s working on an advanced engine with very small cylinders and high bmep (brake mean effective pressure), but wisely abandoned it.

Antoinette (FRANCE)

This was the supreme aero engine prior to the appearance of the Gnome. It was one of the creations of the gifted Léon Levasseur, who named it for his patron's pretty daughter (though Antoinette Gastambide did not do as well as another French girl whose name was Mercédés!). First run as a motor-boat engine in 1905, it was adapted for aero use in 1906. It was a 90° V-8 with cylinders 80 mm square (3.2 litres), cast separately with evaporative steam cooling using electroformed copper jackets. Left and right cylinders were offset to allow two big ends to run on each crankpin. A major feature was direct injection of metered doses of carefully filtered fuel into each of the automatic inlet valves. The crankshaft had gears or belts to drive the injection pumps, oil pumps and the magneto feeding the sparking plugs. Typical weight was 50 kg and rating 24 hp.

In 1908 a 50-hp model with cylinders 110 × 105 mm was sold at £480, compared with £328 for the original engine. By 1909 heads and cylinder barrels were machined from a single steel forging, this 60-hp model having a booster trembler-coil and usually a flywheel. An impressive V-16 gave 100 hp, but the author does not believe the amazing 32-cylinder Antoinette ever flew.

Anzani (FRANCE)

Motorcycle builder Alessandro Anzani of Cour-bevoie sprang into eminence in 1909 with one of his

Above *The classic Antoinette V-8 can be considered the first refined, factory-built aero-engine. This example has an exceptional group of auxiliaries.*

Below *Most powerful of the commoner pre-war Anzanis was the 10-cylinder of 1913, rated at 100 hp. It is not immediately obvious that the cylinders are in front and rear rows. The carburettor is at the bottom, and to reduce oiling-up the spark plugs were inserted on the upper sides of the cylinders.*

first aero engines. A 3-cylinder W or fan type (a V with an extra cylinder vertically upright) with crude air cooling and make-and-break ignition, it had a 105 mm bore and a 130 mm stroke, and a capacity of 3.75 litres. Rating was 24 hp at 1,600 rpm, for a bare weight of 66 kg including a 26-kg flywheel. Provided it did not overheat it was fairly reliable, and one took Blériot to Dover on 25 July 1909. From April 1909 Anzani also sold the engine with 120 × 130 mm cylinders (35 hp) and 135 × 150 mm cylinders (45 hp). By December 1909 he was running a 3-cylinder radial, with 120° between the cylinders. About three months later he had created the first 2-row radial by adding a second 120° 3-cylinder row behind the first, rotated 60° to give all cylinders equal cooling. He retained sprung auto inlet valves, and decided to use a kinked crank-web to bring the front and rear cylinders closer together, using very slim but broad connecting rods and driving via more refined slipper-type big ends. Using slightly smaller (90 × 120 mm) cylinders, the first of these 6-cylinder radials was advertised as giving 45 hp at 1,300 rpm.

By late 1913 Anzani had no fewer than 7 types of aero engine on sale, though they were hand-built in ones and twos. Biggest were the 10-cylinder radials; one of the 110-hp size (105 × 140 mm cylinders) being exhaustively tested at Farnborough in 1914. Anzani production in World War I was mainly for trainers, but before the end of that conflict he had introduced push-rod operation of the inlet valves, with the exhaust valves in line directly ahead. There followed a further profusion of different types, some with water cooling, but none received the concentrated effort needed for high reliability. Small numbers continued to find buyers, one of them Clyde Cessna whose first aircraft had the 120-hp 10-cylinder Anzani engine. In 1928 Potez used Anzani cylinders in the first of his 6-cylinder single-row radials.

Argus (GERMANY)

This company, for most of its life centred at Berlin-Reinickendorf, built its first car engine in 1902, an upright water-cooled 4-in-line, and followed it with a derived aero engine in 1906. Using cylinders initially measuring 124 × 130 mm, and then from 1909, 140 mm square, it was the chief German aero-engine firm until 1914, delivering 490 engines, over a dozen of which powered pre-war 4-engined Sikorskys. Most went into Taubes and derived types. Continuing mainly at the 100-hp level, some thousands were delivered by 1918, by which time a few Argus engines had been delivered at 190 hp. Aero work ceased in 1919, but briefly returned in 1926 with two large 43.5-litre water-cooled V-12 engines of 1,300 hp, one upright and the other inverted. Only two or three of each were made.

In 1928 Argus found its feet with the As 8, an inverted 4-inline with air-cooled cylinders of 120 × 140 mm, 6.3 litres, rated 110 hp at 2,100 rpm. It put the aero side of the company on a firm footing at last, and led to the corresponding inverted V-8, the As 10 of

Below *The Argus As 411 was the last of the company's range of inverted-V air-cooled engines. Renault, despite claims of redesign, did little more than change the accessories and nameplate.*

Bottom *This Argus was made by Opel in 1917. Its 3 pairs of cylinders were 145 × 160 mm and the rating was 180 hp at 1,400 rpm. (Photographed at the Science Museum, London.)*

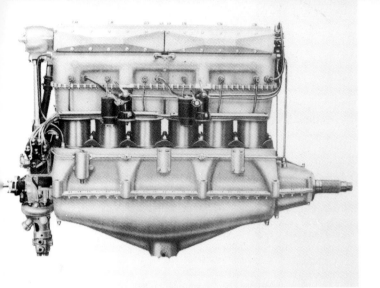

1931. This was rated at 240 hp in later models, notably the As 10 C, of which 28,700 were delivered by 1945. In 1937 the firm developed a smaller cylinder, 105 × 115 mm, matched to greatly increased operating speeds, with deep-finned steel barrels and aluminium heads, and crankcases made from cast Elektron (magnesium alloy). The first production application of this advanced technology was the As 410 inverted V-12, which in geared and supercharged form weighed around 315 kg and was rated at up to 485 hp at 3,100 rpm. Production amounted to 20,900 units for such types as the Fw 189 and Ar 96B, but the final model, the As 411, was built in modest numbers (c 2,600) by Renault in Paris, mainly for the Si 204D. The 411 was a highly rated engine weighing 385 kg as a complete power egg, giving 600 hp at 3,300 rpm. After the liberation Renault continued development.

Argus Motorengesellschaft's most notorious engine was the Type 109-014 Argus-Schmidt pulsejet used to power the V-1 flying bomb and various last-ditch manned aircraft of 1944-45. Made of wrapped mild-steel sheet, it contained a spring flap-valve grid in the inlet which, in operation, set up combustion resonance at about 45 Hz, giving some 272 kg thrust, equivalent at bomb cruise speed to 650-750 hp (bombs flew at various speeds depending on the fuel metering unit). The first, smaller, Argus-Schmidt pulsejet was tested under a Go 145 (itself Argus-powered) in April 1941. The actual bomb first flew on Christmas Eve 1942. Production of the 109-014 amounted to 31,100.

Armstrong Siddeley (UNITED KINGDOM)

This famous Coventry company got started in aero engines when the Royal Aircraft Factory was forbidden to build production engines from February 1917. A month before this Major F.M. Green left the Factory to join the Siddeley-Deasey Motor Car Company as chief aeronautical engineer, taking with him the RAF.8. His prime task, however, was to help Siddeley redesign the BHP, and he assisted chief designer F.R. Smith to turn this engine into the

Top left *Built in large numbers, the Siddley Puma survived longer after World War 1 than it should, especially in airline service. It was at least a big improvement over the Beardmore and BHP.*

Above left *The Armstrong Siddeley Jaguar was possibly the world's best aero engine in the years immediately after World War 1. Diameter was only 43 in, even with the increased 5.5-in stroke. This is a late Mk IVC with enclosed valve gear.*

Left *Armstrong Siddeley Motors not only had an autocratic non-engineer head but also lacked the drive to build competitive high-power engines. The Lynx, a staple product for 19 years, suited their style perfectly.*

Right *Ten is an uncommon number of cylinders for an aviation engine, but the Armstrong Siddeley Serval was virtually two Mongoose on one crankcase. Ratings varied from 295 to 340 hp.*

Below right *There was nothing slipshod about Armstrong Siddeley quality control, but their big engines lacked both power and reliability. This Tiger IX of 850 hp was replaced in the A. W. Ensign airliners by Wright Cyclones.*

Bottom right *Last of the ASM piston engines to fly, the Deerhound was compact and showed promise, but was simply not sufficiently powerful to be worth producing. As in the Boarhound, each row of cylinders was in line.*

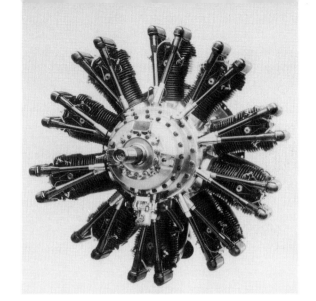

Siddeley Puma, a refined 6-in-line with a cast aluminium block with open-ended steel liners and a separate bolted-on block for the heads. The Puma was built in larger numbers than any other British engine of the day, as described under BHP.

By 1918 pressure on getting the Puma right had eased, and work began in earnest on the RAF.8, which Siddeley named the Jaguar. This compact 14-cylinder 2-row radial was probably the best aero engine in the world from 1919 until at least 1922. It incorporated all the knowledge and philosophy of Green and S.D. Heron, the pioneers of the modern air-cooled cylinder. By 1919 the Jaguar was running with close-finned steel cylinders with aluminium heads attached by short 'quick-threads' and housing two valves each inclined at almost 45°. Siddeley eliminated the supercharger, but after 1920 consented to a direct-drive blower to improve mixture distribution. It had cylinders 5 × 5 in, and gave about 300 hp with fair reliability. By 1921 stroke was increased to 5.5 in, giving a capacity of 1,512 cu in, and the Jaguar then became top engine in Britain at around 400 hp for such aircraft as the Flycatcher, Siskin and Argosy transport. In 1925 it became the first engine in production with a geared supercharger, Green and Heron having invented the necessary centrifugal slipping clutch back in 1918. After 1926 it was surpassed by the Jupiter and Wasp.

Siddeley's Tiger, a fine-looking V-12 with Puma cylinder blocks, ran on the bench in early 1920 at 486 hp and powered the first Sinaia bomber in June 1921. In 1920 a mere test rig with two opposed cylinders did so well the Air Ministry ordered a small production run as the 40-hp Ounce for ultralights. In the same year the prototype of the Lynx, effectively one half of a Jaguar, began testing, and from 1923 it was made in increasing numbers at around 200 hp with hardly any need for development. It avoided the Jaguar's weakness, the two-throw crankshaft, and over 6,000 were built by 1939, a good share of them for the Avro 504N trainer and its successors. Income from the Lynx sustained the company through lean years caused not only by the economic situation and competition from Bristol but, to an increasing extent, the refusal of Siddeley (later Lord Kenilworth) to understand that his often good flashes of inspiration needed to be backed up by solid development.

In the absence of this, ASM (Armstrong Siddeley Motors) was doomed never to be in the front rank, despite the talent of chief designer Harry Cantrill. Engines appeared a-plenty: the Genet emerged in 1925 (five 4 × 4 in cylinders, 82 hp), the Mongoose in 1926 (five 5 × 5.5 in Jaguar cylinders, 150 hp), the Civet in 1927, Double Mongoose (later named Serval) in May 1928, the Genet Major in 1928, the massive 14-cylinder Leopard (6 × 7.5 in cylinders, 800 hp) in 1928, the Panther (14 cylinders 5.3 × 5.5 in, starting at 525 hp) in 1929, the 250-hp Lynx Major in 1929, the Cheetah (7 cylinders 5.25 × 5.5 in) in 1930, and the Jaguar Major (later renamed Tiger, 14 cylinders 5.5 × 6 in) by 1931. The most important of these was the Cheetah, which suited the company's style and capability. Starting at 295 hp, it went on to give 425 hp in the last Ansons in 1952, by which time over 37,200 had been delivered. This filled the Parkside works and took the pressure off the big engines. From the original Jaguar all these had suffered from crankshaft deflections that cried out for a centre bearing. The Tiger gave 800 hp, and was coaxed to 920 hp by 1938 for Whitleys and Ensigns, but reliability was poor; the former aircraft switched to the Merlin and the underpowered Ensign to the Cyclone.

The final fling comprised three superficially attractive 3-row radials, starting with the Hyena flown in 1933. In 1934 Colonel L.F.R. Fell came from Rolls-Royce to be chief engineer, and with Cantrill he planned not only the interchangeability between contrasting types of engine—demonstrated, for example, on the Hampden/Hereford, Wellington, Beaufighter, Halifax and Lancaster—but also the final two completely new high-power engines. The Deerhound and Boarhound were again 3-row radials but owed little to the more traditional Hyena. Extremely compact, the Deerhound had 21 air-cooled cylinders with each line of 3 served by an overhead camshaft. The prototype gave 1,115 hp on test in 1936, and in 1938 it flew in a Whitley II, the final cowling looking like that of a modern C-130. The Boarhound, a bigger 2,250-hp engine, never flew.

Gas turbines

By 1940 ASM had 'got its feet wet' in the challenging new field of gas turbines. In 1939-41 it built on Ministry contract the contra-rotating contraflow gas turbine suggested by A.A.Griffith at the RAE ten years earlier. Thus, when ASM were given their first gas-turbine contract on 7 November 1942 the company chose to go for an axial machine, and finally picked the ASX, a reverse-flow 14-stage axial with 90° elbows at the front feeding 11 slim combustion chambers discharging through a 2-stage turbine. The

ASX ran in April 1943, flew in the Universal Testbed Lancaster and by 1944 was giving 2,800 lb thrust at 8,000 rpm for a weight of 1,900 lb.

This was not competitive, so a propeller gearbox was added to produce the ASP, run in April 1945. This gave 3,600 shp plus 1,100 lb thrust for the rather ponderous weight of 3,450 lb. It was inferior in almost all respects to the Rolls-Royce Clyde, but Hives declined to continue the latter so ASM got a production contract for the ASP, renamed Python. Tested in the outer positions of Lancaster *TW911* and Lincoln *RF403,* the Python and its 8-blade Rotol contraprop suffered prolonged development and lagged the Clyde by more than four years, but at last entered service with the Royal Navy in May 1953 as the powerplant of the Wyvern S.4. The Python 3 had an airflow of 52.5 lb/s, pr of 5.35, weight of 3,505 lb and take-off rating of 4,110 ehp.

Under technical director Bill Saxton and chief designer A. Thomas, ASM went flat-out for axial gas turbines. A contract was received in mid-1945 for a unit in the 1,000-hp class and the prototype Mamba ran in April 1946. At first it was judged a better bet than the rival Dart, but the latter was picked by George Edwards for the Viscount. Mambas flew from 14 October 1947 in Lancasters, a Dakota and the Marathon 2, but found few applications (Apollo and Seamew). Far more important was the Double Mamba, comprising two Mambas, each driving a 4-blade propeller but sharing a common inlet casing and installed as a single powerplant, the propellers being a single coaxial unit. The advantages were that one engine and propeller could be shut down for long-range cruise, and it could burn diesel or other fuel already on board warships, eliminating high-octane gasoline. The ASMD.1 first flew in the Fairey GR.17 on 29 September 1949, this leading to the very successful Gannet ASW and AEW aircraft, the final AEW.3 version having the Bristol Siddeley Mk 102 engine of 3,875 ehp.

An extremely important feature introduced hesitantly with the original ASX was the use of vaporizing

Top right *The Armstrong Siddeley Double Mamba not only pioneered the use of two power sections joined to one gearbox but each half drove its own propeller as in the earlier Fairey Prince. This is a 3,035 ehp Mk 101 (Gannet AS.4 powerplant).*

Above right *An Armstrong Siddeley (later BS, later still RR) Viper 11 of 2,500 lb rating. Vipers went into production at 1,640 lb (as a 'short life' engine) and remain in production 32 years later at 5,000 lb!*

Right *Despite its initial superiority to the rival Avon, the Armstrong Siddeley Sapphire was ordered in much smaller numbers (if one excludes US Wright J65 production). Last variant was this Mk 203/204 for the Javelin 7 and 9, with a low-augmentation afterburner boosting thrust to 13,390 lb.*

combustion. There are inherent problems with high-pressure atomizing burners, but with the ASM system the fuel is sprayed at low pressure into a 'walking-stick' 180°-curved tube where it vaporizes, as in an ordinary blowlamp, to give near-perfect burning at all fuel flows, which can vary 100-fold between sea-level take-off and high-altitude flight idle. When Dr Stanley Hooker became technical director of the combined Bristol Siddeley firm in 1959 he found the ASM system superior to the more common scheme, and ordered it used on former Bristol engines; he would have had it on the RB.211 at Rolls-Royce but this engine was already planned with the traditional arrangement. Vaporizing combustors were on all ASM gas turbines from 1943.

Removing the gearbox from the Mamba in November 1948 gave the Adder turbojet, rated at 1,050 lb thrust for a weight of 580 lb and used in the Pika and Saab 210, and for ASM's first afterburning experiments on 1 January 1950. Saxton and W.H. Lindsey now had enough experience to start afresh, and quickly designed the Viper, a simple short-life turbojet inteded for the Jindivik target drone. The Viper had a new 11-stage compressor (32 lb/s, pr 4), 24-burner vaporizing annular combustor and a variety of external or integral starting systems. Rated at 1,640 lb for a weight of 365 lb, it set entirely new standards. Soon a 'long life' version was flying in the Midge and Jet Provost, and under chief engineer John Marlow its development continued through Bristol Siddeley and Rolls-Royce right up to the present, at ratings up to 5,750 lb and with sales close to 6,000.

Under Sid Allen ASM thrust ahead in 1946 into liquid rocket engines. Funding was slow until the Korean War in 1950; then the 2,000-lb Snarler

quickly appeared, pump-fed on lox/alcohol, followed by the 8,000-lb Screamer for the Avro 720, but all such work was stopped by the 1957 Defence White Paper that terminated new fighter projects. It also cancelled the Avro 730 canard bomber/reconnaissance aircraft designed to fly at Mach 2.5 on P.176 afterburning turbojets in two 4-engine box nacelles.

By far the most important of ASM's engines was the Sapphire, handed to the company in 1947 when Metrovick was told to stop aero work. The former F.9 was a superb engine, and in particular its 13-stage compressor was the best in the world when ASM began bench running on 1 October 1948. Other features were an annular combustor with 36 walking-stick burners, and a 2-stage turbine. An ASSa.3 was type-tested at 7,500 lb in November 1951, and an ASSa.6 passed the test at 8,300 lb in May 1952. Weight was 2,550 lb. Subsequently 200-series Sapphires were cleared at dry thrusts exceeding 11,000 lb, and with a simple afterburner at 13,390 lb for the Javelin. Wright licensed a US version designated J65 (*qv*).

The last ASM project to run was the P.181, a completely new engine in the 1,000-2,000-hp class for aeroplanes or helicopters. The HPR.5, the rebuilt Mamba-Marathon 2, was being readied to fly it in 1958 when the project was abandoned. The company was merged with Bristol Aero-Engines into Bristol Siddeley in 1959.

Arsenal (FRANCE)

The Arsenal de l'Aéronautique et Châtillon-sous-Bagneux had many fine engineers, some of whom wasted years developing flexible couplings for tandem engines. From 1944 it improved the Jumo 213 as the Arsenal 12H, rated at 2,250 hp with water injection. These saw active service in the Nord 1402 Noroit amphibian. The unwieldy 12H-Tandem comprised two of these engines linked by the Arsenal coupling, giving 4,500 hp and intended to replace 12Z engines in Arsenal's own VB 10 fighter. The culmination was the monster 24H, with four 12H cylinder blocks in vertical opposed pairs driving two crankshafts geared to a common propeller shaft. A cross-shaft drove two superchargers, that on the left feeding the upper blocks and that on the right the lower. The 24H weighed 1,900 kg and was rated at

Left *The Arsenal 24H with four Jumo 213 cylinder blocks was only half of a vast tandem engine with 140 litres and 8,000 hp. Even at the time this appeared to be a colossal solution to a non-existent problem.*

Right *The final wartime Austro-Daimler was rated at up to 225 hp. Note the front drive shaft to the overhead camshaft, with one of the 6-cylinder magnetos beside it, and the carburettor group amidships.*

4,000 hp. The propeller drive was arranged to allow a shaft to pass through from a second 24H in the rear; the 24H-Tandem (140 litres) was actually run. The 24H was tested in the inner positions of a Languedoc, driving tiny 5-blade Rotols unable to transmit even half the power.

Austro-Daimler (AUSTRIA)

The first engines designed by Dr-Ing Ferdinand Porsche included an outstanding 6-in-line of late 1910. It was possibly the first of the vast numbers of such engines used in almost all the Central Powers' aircraft of 1914-18 to have been designed from the start for aero use. By 1911 cylinder size had settled at 130 × 175 mm, giving a capacity of 13.9 litres; rating was almost always 120 bhp at 1,200 rpm, and dry weight about 260 kg. Advanced features included welded steel (originally copper) water jackets, seven main bearings and inclined inlet/exhaust valves of large diameter operated by push-pull rods.

From the start the modest rpm and general stress levels resulted in good reliability, and it was competition from this engine that spurred Mercedes, Benz, Hiero and others to enter the aero field. One of the first foreign customers was S.F. Cody in Britain, in April 1911. The same engine had already flown 15 hours and crashed twice when it was the sole reason for Cody's *Cathedral* winning the 1912 Military Trials at Larkhill. In 1911-14 the Austro-Daimler was one of the best-selling engines, the original size being partnered by a smaller-capacity model of 90 hp.

The 1913 Austro-Daimler was built in fair numbers for second-line aircraft during the war, and in Britain served as prototype for the first Beardmore. By 1916 the chief production engine was rated at 185 hp, and in December of that year it was cleared to run at 1,350 rpm giving 200 hp, being used in many Austrian-Aviatik, Hansa-Brandenburg and Phönix aircraft. By late 1917 the power had risen to 210 hp and finally to 225 hp, without change in cylinder size.

Avco Lycoming (USA)
See Lycoming.

B

Beardmore (UNITED KINGDOM)

William Beardmore, of Dalmuir, Scotland, a famed engineering firm, obtained a licence to build the Austro-Daimler engine in 120-hp size not later than March 1914, and quickly introduced a number of changes including dual ignition and twin carburettors. Several thousand were built, terminating in 1917. In 1915, by which time a subsidiary, Beardmore Aero Engine Company, had been registered, the company felt able to effect a more complete revision which included increasing bore to 143 mm

A typical Beardmore 6-in-line, in London's Science Museum.

A Bentley BR.2 with one cylinder cut open for display purposes. This BR.2 was made by Gwynnes Ltd, probably for a Snipe.

giving a capacity of 16.4 litres; take-off rating went up to 160 hp, later cleared to 192 hp at 1,450 rpm, and price from £825 to £1,045. The bigger engine, weighing 592 lb, came into production in March 1916. Like the 120-hp, most of the early output had been reserved by The Royal Aircraft Factory, mainly for the FE.2b. Historian Jack Bruce comments that it was not as reliable as the lower-powered engine. F.B. Halford was instructed to develop the 160-hp version to give greater power, resulting in the BHP series (*qv*). One version of this engine was built by Galloway. In 1919-22 Beardmore worked on the 840-hp Cyclone, a giant 6-in-line of 8.562-in bore and 12-in stroke!

When the British government at last decided in 1924 to build giant commercial airships it decided to allocate one to private enterprise and the other to a government organization. The latter ship, *R-101*, was created on the basis that money was no object, and special giant 4-stroke diesel engines, later named Tornado, were ordered from Beardmore. Each was an 8-in-line with cylinders 8.27 × 12 in, rated at 585 hp at 900 rpm for a weight of 4,225 lb. The massive weight was accepted as being less than the saving in weight of fuel on long journeys, but the *R-101* crashed (having run more than three years behind schedule) with her engines still unable to give their design power and liable to break down at any moment. A distant relative was the Simoon diesel of 1,000 hp, first run in 1923 and flown in the second Blackburn Cubaroo.

Bentley (UNITED KINGDOM)

Already an established motor engineer, Lieutenant W.O. Bentley, RNAS, was sent by Commander Briggs at the Admiralty to study the overheating problem in rotary engines in late 1914. Predictably he decided the best long-term answer was a new engine, and starting with the very expensive Clerget produced an engine with aluminium-alloy pistons designated AR.1, for 'Admiralty Rotary'. It was greatly improved from the manufacturing viewpoint, price being reduced from some £950 to £605, while power was increased to 150 hp, largely by increasing stroke to 6.7 in (capacity 1,055 cu in). By 1915 he was completely redesigning the cylinder, with an aluminium barrel, cast-iron liner and steel head tied to the crankcase by four long bolts, with two valves operated by push rods. This is believed to have been the first production air-cooled cylinder with an aluminium barrel. Bentley also introduced dual ignition. Capacity was the same as the AR.1, and weight increased still further to about 397 lb, but the engine's efficiency and reliability were transformed, the rating of 150 hp at 1,250 rpm being reliably obtained without deterioration over at least 100 hours. Designated BR.1, for Bentley Rotary, this engine was mass-produced for the Admiralty by several contractors, and used to power Sopwith Camels among other types. (Despite this, the Clerget continued to be made under licence in even greater numbers.)

During 1916 Bentley, still a Lieutenant, completed the design of a completely new and challenging rotary using the same principles but with larger cylinders 5.5 × 7.1 in, giving a capacity of 1,522 cu in. Three prototypes were ordered in April 1917. Cleared to 1,300 rpm, the first BR.2 gave 234 hp on its first run in early October 1917, yet weighed only 93 lb more than the BR.1 at 490 lb. The BR.2 was an instant success, and Major-General Sefton Brancker exclaimed that it could with advantage 'be put into every type of aeroplane in France, except bombers'. A production programme was organized for 1,500 per month, the eventual price of £880 still being well under that for the Clerget of half the power. The BR.2 represented the pinnacle of rotary engine development, and in the Snipe continued in RAF service until November 1926.

Benz (GERMANY)

Benz & Cie, of Mannheim, was the pioneer of the motor car from 1885. A 4-in-line water-cooled engine of 100 hp powered the RWMG non-rigid airship of 1909, and several types of derived engine powered aeroplanes and did well in the Kaiserpreis of 1913. In that year the Bz II was put into limited production at a rating of 100 hp, with cast-iron cylinders each with 2

Weighing 275 kg and rated at 180-185 hp the Benz IIIa was one of the most important German World War 1 engines. The propeller flange is on the left, the centrifugal water pump high on the right, and the twin updraught carburettors can be seen feeding the 2 distribution manifolds.

valves 180° apart worked by pushrods from camshafts on both sides of the steel crankcase. The capacity was 14.34 litres, and operating speed was 1,200 rpm. In 1914 the Bz III introduced a higher compression ratio and ran at 1,400 rpm to give 160 hp. The Bz IIIa, the first mass-production Benz, increased capacity to 17.5 litres (cylinders 140 × 190 mm) to give 180 hp, while the final mass-production engine, the Bz IV, increased the bore to 145 mm and was rated at between 200 and 230 hp, though usually called the '200-hp'. More than other engine firms in the Central Powers, Benz dabbled in V-form engines. An oddity was the Bz IIIb of 1918, a 60° V-8 with small 135 × 135 mm cylinders, because even running at 1,750-1,800 rpm this gave only 200 hp. In contrast the Bz VI was a big V-12—effectively two Bz IVs on a common crankcase but with various changes to give a take-off power of 550 hp, more than any other German wartime engine. Two of these, with two Bz IVs, powered the final Zeppelin Staaken Giant, the *R.XVI*.

BHP (UNITED KINGDOM)

In 1916 Frank B. Halford, by this time a Captain in the AID (Aeronautical Inspection Directorate) at Farnborough, was sent to the Beardmore company to develop their 6-in-line (Austro-Daimler) to give 200

hp. He set up an experimental shop in the nearby Arrol-Johnston Car works at Dumfries, and within 6 months was running the prototype BHP (Beardmore-Halford-Pullinger, but having the additional connotation of brake horsepower). Larger than the Beardmore, it had cylinders 5.7 × 7.5 in, giving capacity of 1,148 cu in. Each was made up of a closed-end steel liner threaded over its whole length (Halford copied the Hispano) into a massive cast aluminium barrel to which the cast-iron poultice-type hemispherical head was bolted! On the outside was a welded steel water jacket terminating well above the crankcase. Weight was 690 lb, and rating 236 hp at 1,400 rpm. The prototype ran in June 1916, and despite the several unsatisfactory features showed fair reliability. The same engine was installed in the prototype Airco DH.4 which flew on 5 August 1916. The 42.91-in height of the BHP resulted in an unsightly step in the fuselage top line, the aircraft having been designed for the Beardmore. Flight test results were good, but the BHP needed modifications before it could go into economic production. Various changes, all of a mechanical nature concerned not with performance but with facilitating manufacture, were organized at the Siddeley company and took until spring 1917 to complete.

By this time the BHP was being built by both Galloway Engineering and by Siddeley. The first batch of BHP engines was received by Airco for the DH.4 in July 1917; it was then found that the engine mountings had been changed, unknown to Airco, and the aircraft production line had to be altered. The DH.4 was powered by almost every high-power water-cooled engine on the Allied side, but at least 220 had the BHP in a neat installation with a straight top line and close rounded cowling. Only a handful of Galloway Adriatics were delivered, one being fitted to the prototype DH.9 in July 1917.

The DH.9 had been planned as the standard tactical bomber of the RFC, and its selected engine was the BHP. Such was the importance attached to the programme that in January 1917 Major F.M. Green was detached from Farnborough to serve as chief engineer (aero) of Siddeley-Deasy, which was tooling up to build the engine at ten times the previous rate of 25 per week. An order was placed for 2,000, to be completed by October 1917, by which time more were to be on order. Green's brief was to improve reliability and ease of manufacture, with minimum delay. He had no instructions to increase power, but was confident 300 hp could be obtained and this may have led to the exciting performance estimates for the DH.9. John Siddeley himself had no technical training but did have flashes of inspiration, and against the advice of his own designer, the famed S.D. Heron, he insisted that the BHP cylinder be redesigned com-

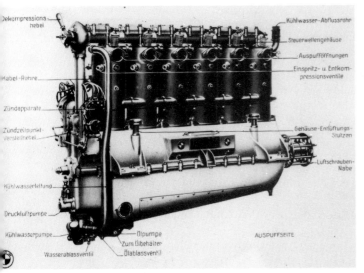

Labels on image:
Dekompressionshebel
Kabel-Rohre
Zündapparate
Zündzeitpunktverstellhebel
Kühlwasserleitung
Druckluftpumpe
Kühlwasserpumpe
Wasserablassventil
Ölpumpe
Zum Ölbehälter
Ölablassventil
AUSPUFFSEITE
Kühlwasser-Abflussrohr
Steuerwellengehäuse
Auspufföffnungen
Einspritz- u. Entkompressionsventile
Gehäuse-Entlüftungsstutzen
Luftschrauben-Nabe

Above *Designed near the end of World War 1, the BMW IIIa was a fine modern engine that led to many others. Note the twin 6-cylinder magnetos, the vertical cooling water pipe from the pump under the rear of the crankcase, and the decompression lever at the rear of the camshaft drive box. (Original BMW training manual.)*

Below *One of the most important engines of the inter-war period, the BMW VI was further developed in the Soviet Union as the M-17 and made in vast numbers. The Munich-built engines played a key role in Hitler's first generation of warplanes.*

pletely. The result was an open-ended steel liner with a short thread at the upper end for attaching the cast aluminium head. This arrangement, also used on the contemporary Jaguar, was revolutionary but gradually became universal for poppet-valve engines. More serious, the welded water jacket was replaced by a single cast aluminium cylinder block. The redesigned engine was named the Siddeley Puma. Production began as early as October 1916, at a rating of 300 hp, the weight being reduced to 625 lb. Most unfortunately, once engines were in mass production in January 1917 it was found that exhaust valves were burning out rapidly and that 90 per cent of the cylinder blocks were riddled with porosity or machining flaws. All that could be done was to derate the Puma back to 230-240 hp, and even then the faulty raw material caused numerous failures. A few DH.9s had the fully rated engine, officially named 290-hp Puma (High Compression), but most suffered with an engine which had become obsolete, deficient in power and inherently unreliable.

BMW (GERMANY)

The Bayerische Motorenwerke AG (Bavarian Motor Works) was founded at Munich-Allach in 1916, and immediately began designing yet another water-cooled 6-in-line. This progressed through 'I' and 'II' stages, with slight variations, and entered production

in February 1917 as the BMW III, with steel cylinders with welded water jackets with bore and stroke 150 × 180 mm and capacity 19.08 litres. Almost all the production engines were deliberately designed with high compression, typically 6.4, taking advantage of the relatively high octane number (not then numerically understood) of the benzole-blended fuel. Run at full power at sea level the engine would have blown apart, so the throttle was gated and only opened progressively as the aircraft climbed, reaching fully open at 2 km (6,560 ft) or more. Often the pilot had three throttles, one being opened fully for take-off and the other two opened slowly in sequence. Take-off power varied between the usual figure of 185 hp for the IIIa and 220 hp in strengthened versions, in each case at a modest 1,400 rpm.

After the war the BMW company was able to keep its design team, headed by Helmut Sachse, occupied in a series of water-cooled engines which gradually assumed a dominant position in Germany and were licensed to other countries. The BMW IV was a 6-in-line with greater capacity than the III and was typically rated in 1921 at 250 hp. The BMW V of 1926 was a V-12 broadly equal to two IIIa engines, rated at 360 to 420 hp. The BMW VI and VII also appeared in 1926, and were V-12s equivalent to a pair of BMW IVs, with the same large size of cylinder (160 × 190 mm) and capacity 46.95 litres. Both began life at a modest 420-440 hp, but the former was intensively developed and built in increasing numbers for a growing proportion of Germany's civil transports and the first generation of Luftwaffe warplanes. Adopted as the standard V-12 of the Soviet Union in succession to the M-5 Liberty it was designated M-17, and subjected to independent development, the 17F being rated at 715 hp. The BMW engines were distinguished by a 'z' suffix denoting compression ratio, the most important versions being the 660-hp 6z and, for military use, the 7.5z rated at 750 hp. At least 9,200 BMW VI engines were built in Germany, terminating in 1938, with Soviet M-17 output being considerably greater. In addition, the Soviet Mikulin bureau (*qv*) developed the BMW VI further into the monobloc AM-30 series.

At one time BMW was purchased by BFW (Messerschmitt), which changed the name to BMW Flugmotorenbau GmbH. Development broadened in 1927 to embrace air-cooled radials, the first production type being the small BMW X and Xa, with 5 cylinders 90 × 92.5 mm (2.93 litres), rated at 68 hp. In 1929 the company decided it could do better by purchasing a licence from Pratt & Whitney for the Hornet, and by the end of that year the BMW 114 had been developed and run, with cylinders almost identical to the US engine (155.5 × 162 mm, 27.7 litres) but with direct fuel injection. This remained a

The supercharged BMW Lanova 114 V-4 was a 4-stroke diesel of 1938. It was a water-cooled radial, installationally interchangeable with the BMW 132 series.

research and development engine, and in 1937 Dr Schwager's team produced the BMW-Lanova 114 4-stroke diesel version, rated 650 hp at 2,200 rpm, with liquid-cooled cylinders and radiators arranged between them at the sides. Production centred on the refined BMW 132, initially with a float-chamber carburettor. By 1945 total deliveries of 16 production versions exceeded 21,000, the 132A, E and H being the main float-chamber models and the Do, Dc, F,J,K,L,M,N,T,U,W,Y and Z having (usually Bosch) direct injection. Wartime 132T output was assigned to French companies. Most models differed only in propeller and supercharger gear ratios. The take-off powers varied from 725 hp for the A to 970 hp for the wartime K and M.

Prototypes only were made of the BMW 116 of 1932, a neat inverted V-12 with cylinders 130.2 mm square (capacity 20.6 litres), rated at 600 hp. More important was the decision of Hitler's RLM (air ministry) in 1935 to fund two high-power radials, one from BMW and a competitor engine from Siemens (Bramo). The latter company was taken over by BMW in 1939, bringing in valuable engineering talent, but it was under Sachse's direction that the 1935 engine went ahead as the BMW 139. This was two BMW 132s on a single crankcase, giving an

18-cylinder engine of 55.44 litres, which ran in 1938 at the design power of 1,550 hp. The 139 powered the prototypes of the Fw 190 and Do 217, but never entered production because of the promise shown by the more compact and fundamentally newer BMW 801, first run a few months after the 139. With only 14 smaller cylinders (156 mm square, 41.8 litres) the 801 was designed to the same power as the 139. All production versions had a Deckel 14-plunger injection pump and a *Kommandogerät* (control unit) which gave the pilot automatic single-lever control of the entire engine, all variables being governed to their optimum settings.

Most 801s were built as a complete engine-change unit which, for the first time, reduced installed drag below that of an equivalent liquid-cooled engine. The

Left *The BMW 132 was distantly descended from Pratt & Whitney's Hornet. This wartime 132 Dc powered several Luftwaffe types.*

Below *The compact BMW 801 was by far the most important radial in Hitler's Luftwaffe, and generally regarded as an unbeatable engine at low altitudes. This view shows the BMW 801A, with cooling fan and oil return pipes across the top of each cylinder.*

cowl fitted very tightly, and was pressurized by a front fan driven at roughly 3 times propeller speed, the air escaping to the rear through a peripheral slit around which, in most installations, the exhaust pipes were grouped to give ejector thrust. The oil cooler was inside a reverse-flow armoured ring round the front of the cowl. The 801A went into production in 1940 at 1,600 hp for a total powerplant weight of 2,669 lb, and later models included the highly boosted E of 2,000 hp and various turbocharged variants culminating in the 801TQ rated 1,715 hp at 12 km height. Total production exceeded 61,000 engines.

First run in 1940, the BMW 802 was broadly an 18-cylinder 801, with advanced and novel features. Prototype rating was 2,400 hp, but development was abandoned in 1942. By this time design was well advanced on one of the most powerful aviation piston engines, the BMW 803. This had 801-size cylinders arranged in two 14-cylinder radial units back-to-back to give a 28-cylinder 4-bank radial; moreover, the cylinders were liquid-cooled and arranged in in-line pairs, each pair with a common head and camshaft. The front 14-cylinder unit drove the front coaxial propeller, while the drive from the rear unit to the rear propeller was via 7 shafts passing between the forward banks of cylinders. Weight of the prototype was 2,950 kg and rating 4,000 hp, or a continuous output of 2,550 hp at 12 km.

Gas turbines

BMW claimed to have 'begun work on jet propulsion' in 1934. The substance of this claim may refer to the start of research on turbosuperchargers for piston engines under Kurt Loehner. This provided a valuable basis of experience, mostly adverse, on high-temperature turbines. In September 1938 Hans Mauch, of the RLM, visited BMW, Bramo, Daimler-Benz and Junkers to attempt to interest them in gas turbines and jet propulsion. The first two showed interest, BMW because of its turbine experience and Bramo because Ernst Udet was threatening them with closure. Bramo's engineering manager Bruno Bruckmann, and head of research Hermann Oestrich, doubted their ability to produce a powerful and efficient gas turbine. Instead they studied the prospects for increasing aircraft flight Mach number by using a piston engine to drive a multi-blade ducted fan similar to recent ducted fans such as Dowty-Rotol's. A small pattern was flown on an Fw 44 in October 1938, with encouraging results, and in early 1938 an ambitious power unit was tested at DVL Göttingen comprising a Bramo Twin Fafnir driving a large variable-pitch ducted fan with afterburning. This gave results far below calculation, and was dropped when DVL predicted ordinary propeller efficiencies better than 75 per cent at 900 km/h (which are just about being realized in 1985).

Inner and outer combustor flame tubes of a BMW 003A. Design, materials and quality control were just adequate to put large numbers of jets into the sky.

By the start of 1939 Bramo's team reluctantly decided to go flat-out on a turbojet, and Bruckmann's efforts to merge with BMW (to avoid total shut-down) bore fruit, collaboration at the technical level from October 1938 being followed by the acquisition of Bramo in July 1939 with the name BMW Flugmotorenwerke. The original BMW team at Munich-Allach had already launched itself on a two-stage centrifugal turbojet, while the former Bramo team at Berlin-Spandau opted for what seemed the optimum engine, an axial counter-rotating unit smaller and lighter for a given thrust than any axial or centrifugal. In December 1938 this was placed under RLM contract as the 109-002 (all jet/rocket engines were prefaced by 109). To provide experimental data quickly a simpler axial turbojet was also ordered as the 109-003, using Loehner's turbine from the centrifugal engine.

The first BMW 003 ran at Spandau in August 1940, using a 6-stage compressor designed by Encke of the AVA with a pr of 2.77, an annular combustor which held overall diameter to 670 mm, and Loehner's turbine with hollow blades to operate in a gas temperature of 900°C. It was soon realized that this temperature was very optimistic, and that almost every part of the 003 was full of problems. From September 1940 a new 7-stage compressor was designed with NACA blading and 30 per cent greater airflow, as well as a different turbine and fuel burner, of which there were 16. The revised engine, the 003A, reached 550 kg thrust on test in December 1942, compared with the previous 260 kg. Meanwhile, even the original 003 had been coaxed to 440 kg, and two of these were cleared for flight in November 1941 and

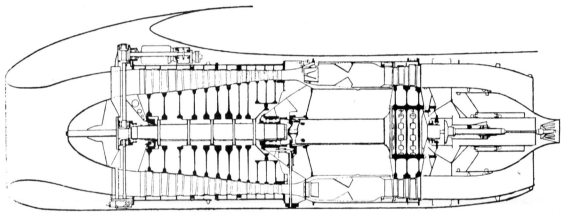

sent to Augsburg to be installed in the Me 262 V1. Still with its Jumo 210 in the nose (a wise precaution) this aircraft flew on 25 March 1942, both BMW turbojets quickly flaming out.

Hans Rosskopf, chief designer at Spandau, completed the drawings for the pre-production 003A-0 in September 1942, and most of the more serious faults were cured in the first half of 1943, though an A-0 did not fly until 17 October of that year, under the nose of a Ju 88A-5. By late 1943 the 003A was lifed at 50 h, giving a take-off thrust of 800 kg (1,760 lb); in the longer term the RLM believed it could equal the thrust of the bigger and more complex Jumo 004B. As it offered greater promise the RLM put enormous effort behind the 003A, deliveries reaching 100 by August 1944 and 1,000 by 27 September. About 3,500 had been produced by the end of 1944, most of them never reaching airframe plants (He 162A production alone was to reach 2,000 per month).

The 003C with a high-efficiency Brown-Boveri compressor ran just once, in May 1945; it was then taken over the border into Austria and hidden in a haystack. A month later Bruckmann visited the spot with Sir Roy Fedden, and the 003C was flown to

Top *When the BMW 109-003 was an active programme there was no time to make a sectioned display exhibit. This 003E-1 was constructed in the late 1950s. The E was the final production version for the He 162, similar to the A-1 but with a 30-second boost rating of 920 kg.*

Above *The BMW 109-018 turbojet was just being assembled as the Allied troops reached Munich-Allach. This was an ancestor of the French Atar.*

Northolt, dismantled without metric tools and taken up in the lift to Fedden's office in Thomas Cook's building in Berkeley Street. As for the redesigned 003D with 8-stage 4.9 pr compressor and 2-stage turbine, this was even less developed. The 003R, an 003A packaged with a BMW 718 rocket engine with sea-level thrust of 1,225 kg, did reach the flight-test stage in the Me 262C-2b. The bi-propellant 718 was dangerously temperamental, but on the one flight test in March 1945 the pilot was afraid the flaps and gear would be torn off before he could complete retraction; the 262 went steeply to 6 km in 50 s, the rocket propellants then being exhausted, and coasted on up to 8 km!

The BMW 109-028 was launched in February

1941 as a turboprop to give 8,000 ehp at 800 km/h at 7.2 km, with 12-stage compressor, 24-burner chamber and 4-stage turbine. For the Ju 287 the corresponding turbojet, the 018, with a 3-stage turbine and rated at 3,400 kg static thrust, was initiated in 1943. The 028 was never completed; the 018 was almost ready for test by December 1944 but was never run because it could be seen to have no influence on the war.

Boeing (USA)

This famous company set up a propulsion department to do gas turbine research in 1943. Soon it was designing hardware, the company assigning type numbers 500-600 for the results. The numbers did not get beyond 502, but No 502 sustained a useful programme which was so diverse the company formed an Industrial Products (later Turbine) division. First to run, in 1946, was the Model 500 turbojet, with a single-sided centrifugal compressor outstanding for its day, with pr of 4.25 at 36,000 rpm, thrust being 180 (later 210) lb, there being 2 combustors and a 1-stage turbine. By adding a free power turbine the Model 502 turboshaft/turboprop was born, initially with an output of 270 hp. This found many customers and developed in power to 300 and later 365 hp in major drone-helicopter applications, with Army/Navy designation T50, other models going up to 500 hp. Output reached 500 by 1956 and about 1,600 at completion in April 1968. Rights were then sold to Steward-Davis Inc. Basically the same engine powers the Swedish S-type tank.

Bramo (GERMANY)

The Siemens-Schuckert Werke, part of the giant Siemens electrical group, produced airships, combat aircraft and numerous rotary engines before and during World War 1. After building licensed Gnome engines in 1911 the Siemens und Halske subsidiary developed the Sh 1 contra-rotating rotary in 1914, the crankshaft rotating one way and the cylinders the other, both being geared to the engine mount and the propeller being on the crankshaft. The Sh 1 had 9 Monosoupape-type cylinders 124 × 140 mm, each half of the engine rotating at 800 rpm, the output being 100 hp. About 180 Sh 1s were built in 1916-17, followed by larger numbers of the Sh 3 and 3a. These had 11 similar cylinders and thus capacity of 18.6 litres. The Sh 3 was rated at 160 hp, each half turning at 900 rpm, and the 3a delivered 240 hp.

By 1921 Siemens had completed prototypes of two sizes of static radial with sound but pedestrian features, using 7 aluminium cylinders with screwed-on heads and steel liners and the 2 valves in line, exhaust at the rear. The most important production engines were the Sh 5 of 1921, with 100 × 120 mm

cylinders and rated at 77 hp, the Sh 11 of the same 6.6-litres size but rated at 96 hp, and the Sh 14 of 1931 with bore increased to 105 mm (7.1 litres) and rated at 113 hp. These led to the Sh 14A of 1934 with bore of 108 mm (7.7 litres) and rpm raised from 1,720 to 2,200 to give 160 hp. This was still in full production when in 1936 the works at Spandau was reorganized as a separate company, Brandenburgische Motorenwerke GmbH, trading as Bramo. The Bramo Sh 14a powered such major types as the Bu 133, Fw 44 and He 72, output certainly exceeding 15,000.

In 1927 Siemens had purchased a licence for the Bristol Jupiter, and this continued in modest production until about 1935. This greatly influenced the first big Siemens engine, the Sh 20 of 1930, from which was developed the production Sh 20B with cylinders 154 × 188 mm, giving capacity of 31.5 litres. This was larger than the Jupiter, because of the increased bore, yet even though stroke was reduced the overall engine diameter was increased; and as maximum rpm were only 1,850 the take-off rating was only 540 hp. Undaunted, Siemens redesigned the Sh 20B to run at 2,500 rpm but with a stroke of 160 mm, reducing capacity to 26.82 litres, giving powers which started at 600 hp in 1933 and reached 760 hp two years later. A new numbering system was adopted, this engine being the SAM 22B, from 1935 called the 322B. It was made in many sub-types for early Luftwaffe use but was always troublesome and unpopular. Incidentally, in 1930 Siemens introduced sintered aluminium oxide spark plugs.

Continued refinement and the introduction of a 2-speed supercharger and Bosch fuel injection resulted

Visitors to London's Science Museum can gently rotate this Siemens Sh 3 (text, see Bramo) and will eventually fathom out how the 11 cylinders go one way and the crankshaft the other!

Above Bramo's Fafnir was taken over by BMW and eventually ousted the parent firm's own Type 132. This is a Fafnir 323AQ of 1,000 hp.

Ministry was anxious to see the Jupiter continue, and applied pressure to the aloof and autocratic board of the Bristol Aeroplane Co to buy the aero-engine assets. For £15,000 Bristol got Roy Fedden and his team of 31 engineers, five Jupiters and a mass of parts, drawings, patterns and tools, and a promised order for ten production engines. Fedden undertook to develop the engine in two years within a budget of £200,000. Sadly, the board did not include a single engineer and regarded Fedden as a mere employee; he, on his part, misunderstood their continued reluctance to put up money for his often optimistic schemes, and from the launch of the Engine Department on 29 July 1920 the lack of communication between Fedden and the board sowed the seeds of serious difficulty 20 years later.

After one year the total outgoings stood at £197,000 and the board decided at its next meeting it would close down the 'cuckoo in the nest' at the end of the

Below One of the later Bristol Jupiters, with each cylinder equipped with a forged head with partially enclosed valve gear.

in the SAM 323, from 1936 called the Bramo 323 and named Fafnir. Cylinders were forged steel with valves at 35° in a head of Y-alloy, crankcase was aluminium and the rear cover and accessory gearbox, magnesium (Elektron). At least 14 sub-types of BMW Bramo 323 were produced in series, to a total of over 5,500, some types such as the Hs 126 and Fw 200C switching over from the BMW 132.

Nevertheless Udet told Bramo their output of about 100 Fafnirs a month did not justify official support, despite the fact that the RLM (Air Ministry) had previously been careful to maintain industrial competition. In 1935 it had ordered a new engine in the 2,000-hp class from Siemens, and when the company heard it might be closed it urgently rushed through prototype testing of this engine, the Twin Fafnir (but with only 7 cylinders in each row) and was delighted to record 2,000 hp, in October 1938. BMW was still a long way from this power. The RLM did not relent, however, and merged Bramo into BMW, killing the Twin Fafnir. Production of the 323 Fafnir continued to 1944, the most powerful, the 323R-2, being rated at 1,200 hp. Bramo's gas turbine history is dealt with under BMW (*qv*).

Bristol (UNITED KINGDOM)

When Cosmos (*qv*) failed in January 1920 the Air

year. But in September 1921 the Jupiter II, with auto-compensation for cylinder expansion on the valve pushrods, became the first engine to pass the severe new Air Ministry type-test, reaching 400 hp at 1,625 rpm, and Fedden obtained permission as a last straw to exhibit the Jupiter at the Paris airshow in October 1921. Here it created intense interest, as the only post-war high-power engine, and the famed Gnome-Rhône company purchased a licence. In December the Air Ministry began formal consideration of the Jupiter as its first new post-war engine for the RAF, and the vital order for 81 engines was signed in September 1923. It was a close-run thing.

From then on, Fedden's insatiable drive forced the Jupiter to the very forefront, culminating after ten years with 17 foreign licensees and a total of more than 7,100 Jupiters for at least 262 different types of aircraft. The Jupiter IV passed the British and French

type-tests at 436 hp in March 1923. In November 1923 the Mk V introduced a split crankshaft with maneton coupling and one-piece master rod with floating big-end. In June 1925 the Mk VI brought in a drop-forged duralumin crankcase and Bristol triplex carburettor, weighing 771 lb and rated at 480 hp. The VII added an outstanding geared supercharger, giving 460 hp. The VIII of 1928 introduced Farman reduction gearing, enabling the engine to run at 2,200 rpm with reduced vibration driving a larger propeller with higher efficiency. Forged screwed/shrunk heads replaced the old poultice head, such engines adding suffix letter F; the IX went to 525 hp at unchanged 5.3 compression ratio; and the supercharged and geared Jupiter X gave 530 hp at 16,000 ft, against some 220 hp at this height of the Cosmos Jupiter I. The HP.42 and other large airliners of 1929 used the Jupiter XFBM, with Bristol gas-starting operated from the cockpit.

Other 1920s engines included the flat-twin Cherub of 1924 of 61 cu in capacity with ball and roller bearings for the crankshaft and big ends, increased in capacity to 74.9 cu in in 1926, raising power from 32

Bristol's neat Cherub had crude finning but enclosed valve gear. Every part of every Bristol engine had an FB (Fedden/Butler) part-number cast or stamped on it.

to 36 hp. The Orion, a turbocharged Jupiter VI, of 1926, was rated at 495 hp at 20,000 ft. The Mercury of 1926 was initially a short-stroke (6.5 in) Jupiter designed for Schneider racing, and giving 960 hp for a weight of 682 lb with diameter reduced from 53 to 47.5 in. The 200-hp Titan of 1927 had 5 Mercury-size cylinders, and the 300-hp Neptune of 1929 used 7 similar cylinders. By 1929 the forged-head Titan was giving 240 hp, but most Titans were licence-built by Gnome-Rhône.

Originally called the Mercury V, the first Pegasus in 1932 restored Jupiter cylinder size (5.75 × 7.5 in) and capacity (1,753 cu in), but was mechanically improved throughout, with more and deeper cooling fins, enclosed valve gear driven by just two pushrods inside an oval tube, lighter reduction gear, and special automatic fuel and oil control systems for rapid take-off from cold with enhanced power. The Draco of 1932, a Pegasus with direct injection, was soon dropped. By 1936 production Pegasus engines were rated at over 800 hp, and by 1939 this had risen to 1,065 hp on 100-PN fuel, for a weight of 1,110 lb. Pegasus production was completed in 1942 at about 17,000, 14,400 of these being delivered after October 1936. The shorter-stroke Mercury also came into production for fighters, the Mk IVA being cleared at 560 hp in 1931, in which year an IVS2 was tested at overspeed of 2,600 rpm giving 893 hp on 77-PN fuel. Fedden was the most important single driving force in the British industry for the variable-pitch propeller

and 100-PN fuel, and the Mercury and Pegasus were the first British engines cleared for their use. The 1,520 cu in Mercury was rated in production at up to 995 hp at 2,750 rpm at 9,250 ft for a weight of 1,000 lb. Total production was 21,993, of which 20,700 came after 1936.

In 1931 a short-stroke (5 in) Mercury in a tight long-chord cowl increased the speed of a Bristol Bullpup 11 per cent compared with a Jupiter but remained a one-off. In 1927 Fedden planned a diesel of 1,000 hp, but this was not built; extensive flight testing was, however, done on the Phoenix, a version of the Pegasus which in 1934 set a diesel height record of 27,453 ft.

By 1926 Fedden was looking ahead to the next generation, and, unable to design an elegant twin-row engine with four valves per cylinder, picked the Burt-McCollum type of sleeve valve. The first research cylinder ran in 1927, and it was soon clear that sleeve valves would be such a stupendous task that an interim next-generation engine was designed. The Hydra, so named because it was double-headed, had 16 small (5 × 5 in) cylinders in eight pairs each with a common head with twin overhead camshafts. In 1933 it flew in a Hawker Harrier (biplane) giving 870 hp at a remarkable 3,620 rpm, but it needed a centre bearing and by this time 870 hp could be seen from the Pegasus and Mercury. Effort on sleeves was redoubled—mass-producing interchangeable sleeves proved a giant task which cost millions of pounds but

Bristol's 16-cylinder Hydra needed a centre bearing, despite its short and stiff crankshaft. Otherwise it was impressive.

The strangely clean appearance of the first Bristol sleeve-valve engines is evident from the Hercules of 1938. Rating was 1,375 hp.

it led to superb engines. With the benefit of hindsight the simpler answer might have been to fall back on just two poppet valves per cylinder and build big two-row engines as did the firm's rivals.

The first complete sleeve-valve engine was the Perseus, the same size (5.75 × 6.5 in) as the Mercury. Run in July 1932, it was type-tested in the same year at up to 638 hp, flew services on Imperial Airways in June 1935, powered RAF No 42 Sqn Vildebeeste IV torpedo bombers in 1937 and was subsequently delivered in quantity at 905 hp. The Aquila, with 9 cylinders 5 × 5.375 in, gave 500 hp in 1934 but did not go into production. The Hercules, with 14 Perseus cylinders giving capacity of 2,360 cu in, first ran in January 1936. The prototype gave 1,290 hp, and development continued on 87-PN fuel at 1,425 hp and 100-PN at 1,590 hp with improved oil-sludge centrifuges, closer-pitch cooling fins, fixed big-end bushes and crankshafts with Salomon vibration dampers. Neville Quinn transformed the previously poor supercharger inlet. Later marks had a steel head into which was shrunk a close-finned copper base for maximum heat emission. By 1945 Hercules production exceeded 57,400, most at 1,650 or 1,735 hp. Post-war military and civil marks gave 1,990 to 2,140 hp, many being licence-made by SNECMA, and the wartime HE.20SM was tested at over 2,500 hp on 125/165 fuel.

Planned as a successor to the Pegasus, the Taurus was run in November 1936. A compact engine with 14 cylinders 5 × 5.625 in, capacity 1,550 cu in, and only 46 in diameter, the Taurus was type-tested in 1938 at 1,065 hp and produced in fair quantity for the Beaufort and Albacore at 1,130 hp. More important was the big Centaurus, with 18 cylinders 5.75 × 7 in, giving capacity of 3,270 cu in. Planned for heavy aircraft, the first Centaurus ran in July 1938 and was type-tested in 1939 at 2,000 hp, but foolishly the importance of this engine was overlooked until 1943, despite the fact that the CE.4S prototype reached 421 mph in a Hawker Tornado in October 1941. In late 1942 the 2-speed supercharged Centaurus was type-tested at 2,375 hp, and cleared for production—in an underground quarry at Corsham—as the Mk V and Mk XI at 2,520 hp. Many other versions followed for military and civil aircraft, most having a Bendix or Hobson/RAE injection carburettor and being rated at up to 2,810 hp with water/methanol injection. Post-war versions included the 2,625-hp Mk 661 for the Ambassador, 2,940-hp Mk 173 for the Beverley and 3,220-hp Mk 373 with direct injection into the cylinders. Before Fedden left the company in 1942 he had initiated construction of the Orion, with 18 cylinders 6.25 × 7.5 in, capacity 4,142 cu in, to give 4,000 hp.

Gas turbines

Despite work on turbochargers from 1923, and a visit from Whittle in 1931, Bristol ignored gas turbines

By 1945 the 'Herc' looked more cluttered; this is a 2,040-hp Mk 739 made under licence by SNECMA, mainly for the Noratlas.

Last of the great Bristol piston engines, the Centaurus went into production in this form in 1943 as the 2,520-hp Mk V, with rear exhaust stacks for close cowling and ejector thrust in the Tempest II.

The starboard outer Theseus turboprop of the Hermes 5, with everything opened. Britain was full of great technical achievements where 'the bottom line' was a zero.

until December 1940 when Frank Owner began formal study of a turboprop in the 4,000-hp class to have fuel consumption at least as good as an equivalent piston engine, cruising at 300 mph at 20,000 ft (the lowest speed and height at which a turboprop was considered competitive). Soon this was thought too ambitious and the big engine was scaled down to become the Theseus, a 2,000-hp machine planned with a heat exchanger to increase thermal efficiency. This had a front annular inlet to an 8-stage axial compressor followed by a centrifugal stage delivering via 8 axial pipes to the heat exchanger at the rear; here the flow was turned 180° to pass through 8 combustion chambers arranged between the air pipes, the hot gas again being turned 180° to pass through the 2-stage gas-generator turbine and independent single-stage power turbine driving the propeller gearbox, before going through the heat exchanger to the jetpipe. This complex arrangement resulted in a weight of 2,800 lb, 500 lb of this being the heat exchanger which took hot gas at 500°C through 1,700 stainless-steel tubes and delivered the incoming air at 300°C, saving a calculated 150 lb of fuel per hour.

The heat exchanger was not ready when the Theseus was first run on 18 July 1945; it gave persis-tent trouble when fitted five months later and was abandoned. In December 1946 the Theseus became the first turboprop to be type-tested, and eventually extensive flight development was done in Lincoln and Hermes aircraft, the Hermes 5 being powered by four fully-engineered Theseus power units which impressed by their quiet running.

In September 1944 work began on a totally new 4,000-hp engine, named Proteus. It was planned for use in the Brabazon 2 and Princess, in each case in a coupled installation buried in the wing, fed from inlets in the leading edge. Thus the reverse-flow layout was deemed appropriate and retained, but in this case the air entered the engine towards the rear, passing between the combustion chamber delivery pipes and turning forward to enter the 12-stage axial compressor followed by two successive centrifugal stages, the intended pr being 9, with airflow 40 lb/s at 10,000 rpm. The high-pressure air then travelled back through the 8 slim combustion chambers and out via a 2-stage compressor turbine and an independent single-stage power turbine driving the 11.4-ratio propeller gearbox.

To assist development the gas-generator section was built separately as a turbojet, named Phoebus, and this ran in May 1946 and flew in the bomb bay of a Lincoln. It did not even reach the design thrust of 2,540 lb and it was found that, far from raising pressure, the first centrifugal stage actually reduced it because of poor flow. It was therefore removed and replaced by a carefully profiled diffuser passage, though not before the first Proteus had begun testing on 25 January 1947. Results were very poor, and even after removal of the first centrifugal stage, giving a 5.35 pr, it seemed as if everything that could give trouble did so. Two Proteus 2s flew in a different Lincoln on 12 January 1950.

A year earlier Dr S.G. Hooker had joined the company and became chief engineer. His predecessor Frank Owner said to him 'I decided we should make the engine with the lowest fuel consumption in the world, regardless of weight and bulk. So far we have achieved the weight and bulk'. Hooker totally redesigned the Proteus, among other things using a 2-stage power turbine of much higher efficiency, and the resulting Proteus 3, or 700-series, was 12 in shorter, 1,000 lb lighter, much more fuel-efficient and given power increased from 2,500 to 3,780 shp. In production this was raised to 4,445 ehp.

To Hooker's relief the Brabazon and Princess were cancelled, but they left the engine with the reverse-flow layout which at the eleventh hour, as the Britannia was being delivered to BOAC, caused an irritating flame-extinction problem due to ice accretion. There was never any danger, and the problem could have been avoided by choosing a slightly different cruise height, but BOAC did all it could to magnify it, delayed acceptance two years, wrecked the Britannia sales prospects and very nearly broke Bristol financially. Subsequently the Proteus flew millions of trouble-free hours and pioneered the use of aero gas turbines for warships and electricity generation, respectively in 1958 and 1959. It also pioneered the use of precision-cast turbine blades, used on all subsequent Bristol engines.

Hooker designed a completely new turboprop, the BE.25 Orion, with a 7-stage LP and 5-stage HP compressor, cannular combustor, single-stage HP turbine and 3-stage LP turbine driving the LP spool and propeller. The Orion was flat-rated at 5,150 ehp to 15,000 ft, sea-level power being potentially about 10,000 hp. With specific fuel consumption of 0.39 it was a fine engine, but the market—mesmerized by jets—evaporated.

In 1946 Owner asked for studies of a high-compression turbojet for long-range bombers. The result was the first 2-spool turbojet, with the HP turbine driving an HP compressor (Owner wanted a centrifugal but was persuaded to have an axial) via a hollow shaft through which passed the drive from the LP turbine to the LP compressor. The resulting BE.10 Olympus ran on 16 May 1950 at a design thrust of 9,140 lb; in fact on its first run Hooker took the throttle himself and deliberately banged it wide open to record a full 10,000 lb. From the start the Olympus was a superb engine, the antithesis of the Proteus, and the Olympus-Canberra testbed soon reached the record height of 63,668 ft, followed by 65,876 ft. The Mk 101 for the Vulcan B.1 had a 6-stage LP spool, 8-stage HP, cannular combustor with 10 burners and single-stage HP and LP turbines. It entered production at 11,000 lb in 1955. The Mk 102 added a zero-stage on the LP spool to increase overall pressure ratio from 10.2 to 12, raising thrust to 12,000 lb. The Mk 104 introduced improved blade material which enabled temperature and rpm to be increased, giving 13,500 lb. All Vulcan engines were brought to this standard at overhaul.

If only the Brabazon 2 and Princess had been cancelled earlier, 'Doc' Hooker would not have had to adhere to the reverse-flow layout in the Proteus turboprop. It was that very feature that delayed Britannia services from 1955 to 1957 and almost broke the Bristol company. This is a production Mk 765.

In 1956 Hooker authorized a redesign of the Olympus with much greater airflow (roughly 240 instead of 180 lb/s), whilst maintaining the existing diameter and pr of 12 and yet using only 5 LP stages and 7 HP. This Olympus 6, designed originally for 13,500 lb, began running in late 1956 at 16,000 lb, soon raised to 17,000. It was intended for the G.50 thin-wing Javelin, and competed for the Vulcan B.2, but the Ministry (in its wisdom) refused to support it and instructed Avro to redesign the B.2 for the rival Conway. Sir Reginald Verdon Smith showed typical nerve in offering to develop the new Olympus at company expense and sell it in production at the same price as the Conway. It proved a massive success, entering service as the Mk 201 at 17,000 lb in July 1960, later becoming the Mk 202 with modified bleed systems. A zero-stage was then added to yield the Mk 301 rated at 20,000 lb, which replaced many 201s in Vulcans from May 1963. For later versions see Bristol Siddeley.

Top *Bristol, then Bristol Siddeley and finally Rolls-Royce (but the same team throughout) took the Olympus from 9,140 to 40,000 lb thrust in 20 years. This is a 20,000 lb Mk 301 for a Vulcan B.2.*

Above *The Bristol Orpheus BOr.12 was the most powerful of this very simple family, rated at 6,810 lb, or 8,170 with 'wee-heat' augmentation.*

Among the prototypes were the Janus 500-hp turboprop of 1948, and the 3,000-lb BE.17 expendable turbojet for the Bristol 182 (Red Rapier) cruise missile. This in turn led to the 3,750-lb BE.22 Saturn designed for the Folland Gnat light supersonic fighter. The Ministry never supported these, but Bristol was convinced of the market for a small axial engine and put Bernard Massey—who had spent years working on Princess auxiliary plant—on the design of the BE.26 Orpheus. This had a 7-stage compressor (the same as the LP spool of the Orion) driven by a single-stage turbine via a shaft so large in diameter, with very thin walls, that no centre bearing was needed. Another novel feature was that each of the 7 flame tubes in the cannular combustor incorporated one-seventh of the ring of turbine inlet vanes. The Orpheus ran at 3,000 lb on 17 December 1954 and was soon type tested at 3,285 lb. It was selected for all contenders in a NATO light-fighter contest, won by the Fiat G91, so the Mutual Weapons Development Program (MWDP) paid for most of the development. Subsequently thousands of Orpheus were produced by Bristol Aero-Engines (the separate company formed in January 1956), Fiat, KHD, and Hindustan Aeronautics, mostly as the Mk 703 rated at 4,850 lb or Mk 803 rated at 5,000 lb.

Bristol Siddeley (UNITED KINGDOM)

When the specification for the TSR.2 aircraft was issued in 1958 the Ministry of Supply used it as a 'carrot' to enforce mergers in the aircraft industry. The airframe companies did not join until 1960, and retained separate identities until 1964, but the engine suppliers carried out a complete merger (between Bristol Aero-Engines and Armstrong Siddeley Motors) in late 1959, the resulting company being BSEL (Bristol Siddeley Engines Ltd). From the start it was a good unified team, continuing existing programmes at Bristol and Coventry and developing new engines under technical director Dr Stanley Hooker. In 1961 it absorbed DH Engines and Blackburn Engines, whose airframe parents joined Hawker Siddeley.

The most important immediate development task was the TSR.2 engine, the Olympus 22R, planned for production as the Mk 320. This was basically a Mk 301 restressed for supersonic flight at sea level or over Mach 2 at height, with a large afterburner. An Olympus had run with a Solar afterburner in 1956, but the 320 was to have a more advanced type with a fully modulated nozzle. A high proportion of the new engine had to be titanium, Nimonic or other high-temperature alloys. Bench testing began in March 1961, followed by the definitive Mk 320 with an enlarged turbine section and 40.5-in diameter afterburner. Ratings were 19,610 lb dry and 30,610 lb with full afterburner, and 33,000 lb was soon demonstrated on test. The TSR.2 flew on 27 September 1964, and a production line was being set up when the whole programme was cancelled for political reasons on 6 April 1965.

Fortunately the Olympus was almost ideal as the basis of the propulsion system for the Concorde SST. At first a version known as Mk 591/2 was selected, based on the TSR.2 engine; then the bigger 593/3 followed with improved cooled HP turbine; then the 593D raised dry thrust from 22,700 lb to 29,300 lb by adding a zero-stage to the LP spool and removing one

from the HP to increase airflow; then the 593B raised thrust to a Stage-0 rating of 32,825 lb and Stage-1 of 35,080 lb; production Mk 602 engines were delivered at the latter figure, raised after two years of service in 1977 to the Mk 621 level of 39,940 lb. Airflow is 410 lb/s and static pr 15.5. French partner SNECMA was responsible for the afterburner, with integral variable nozzle, thrust reverser and noise attenuator. Each engine weighs 5,793 lb.

In 1958 Hooker's team began the design of the first vectored-thrust engine for vertical take-off, the BE.53. It stemmed from a very different conception put forward by former French aircraft designer Michel Wibault. As it involved a manned combat aircraft the British government was precluded from showing interest, because such aircraft had been officially pronounced obsolete. Boldly the board of the Hawker company decided to finance two P.1127 prototypes, while for the second time the MWDP put up 75 per cent of the money for the engine, Verdon Smith for BSEL again immediately agreeing to pay the rest. Thus was the Harrier allowed to happen. Take-off thrust was initially 8,000 lb, using an

Totally unlike anything seen before, this Bristol Siddeley BE.53/2 led to today's Pegasus. The two pipes conveyed 100°C bleed air to cool the 650°C rear nozzle bearings.

Orpheus as gas-generator with shaft drive to the first 3 stages of an Olympus LP compressor delivering via curved left/right nozzles which could rotate 90° to give thrust or lift. In the BE.53/2, mirror-image blading enabled the LP and HP spools to contra-rotate to minimise gyroscopic torques. In the Pegasus 1 the spools were close-coupled, with a new 2-stage fan overhung ahead of the front bearing with no inlet guide vanes (then a radical idea), and with the hot jet bifurcated to a second pair of left/right vectored nozzles. This engine ran at Bristol at 9,000 lb in September 1959.

By this time BSEL was being distracted by a giant NATO scheme for Mach-2 V/STOLs which resulted in the urgent development of the BS.100, an advanced vectored turbofan with PCB (plenum-chamber burning—in effect afterburning in the fan nozzles) giving take-off thrust up to 33,000 lb. This would have been a very important and useful engine, and both the RAF and Royal Navy were allowed to consider versions of the Hawker P.1154 to be powered by it, but like TSR.2 these aircraft, and the engine, were all cancelled in early 1965.

The Pegasus was allowed to continue, however, and progressed through various stages to 11,000 lb (P.1127 flight, September 1960), 15,500 lb (Pegasus 5, fitted to Kestrel), 19,000 lb (Pegasus 6) and 20,500 lb (Pegasus 10) to the 21,500 lb Pegasus 11 which powers the first generation of Harriers. These have a 3-stage titanium fan with part-supersonic blades handling an airflow of 432 lb/s, 8-stage HP compressor, again of titanium, with overall pr 14, and an annular combustor with ASM-type low-pressure vaporizing burners. The Pegasus 11 also has a 2-stage HP turbine (both rotor stages air-cooled and with cast first-stage blades), a 2-stage LP turbine, 4 nozzles (front steel, rear Nimonic) driven over an angular range of 98.5° by chains from a pair of air motors, and a packaged gas-turbine starter/APU (auxiliary power unit) mounted on top. Sea Harriers have the marinized Mk 104, the 11-21E in various sub-types is in production in partnership with Pratt & Whitney for the AV-8B/GR.5, and PCB research is once more being undertaken with a view to vectored engines of over 33,000 lb thrust for future supersonic applications.

Much of the development of the Viper (see Armstrong Siddeley) took place under BSEL. For short-haul jets the BE.47, the jet derived from the Bristol Orion, was replaced by the all-new BS.75 of around 7,700 lb thrust, which was run in 1959. BSEL took over the Gnome, Gyron Junior and other engines from de Havilland (*qv*) and the Nimbus from Blackburn. The latter had purchased a licence from Turboméca in 1953 and planned a range of inter-related engines initially based on the Palas/Turmo/

Palouste all with redesigned inlet and accessory systems. In July 1958 Blackburn ran a turbojet based on the Artouste 600 with an added axial zero-stage, leading a month later to the 840-hp A.129 turboshaft from which was refined the Nimbus of 968 hp, flat-rated in helicopters at 710 hp and produced by BSEL.

A GE T64 licence was not taken up, nor were other planned marketing deals and a range of completely new small turboshafts and lift jets, other than the BS.360 which became the Gem (see Rolls-Royce). The BS.605 was an assisted-take-off rocket for South African Buccaneers; resting on extensive ASM and DH rocket experience, it was a 367-lb retractable package pump-fed on HTP (high-test peroxide) and kerosene to give 8,000 lb thrust. In October 1966 BSEL was purchased for £63.6 million by Rolls-Royce, which wished to stave off competition from a planned JT9D link between BSEL, Pratt & Whitney and SNECMA.

Campini (ITALY)

Secondo Campini was a pioneer of jet propulsion, but without the crucial adjunct of the gas turbine. He got Caproni to build an aircraft, the N.1, powered by an Isotta-Fraschini Asso piston engine of 900 hp. This drove a 3-stage fan which blew air from the rear jet nozzle. To boost performance fuel could be burned in the jetpipe. First flight was on 27 August 1940. Performance was worse than if a normal propeller had been used.

CFM International (FRANCE/USA)

From 1968 General Electric and SNECMA independently concluded that a market existed for a modern commercial turbofan in the 10-tonne class; GE worked on the GE13 (the core of which is used in the F101) and SNECMA the M56. In 1971 SNECMA sought a partner and in December reached an agreement with GE. CFM as a joint company was formed in 1974 to manage the programme for the agreed engine, the CFM56. GE assumed responsibility for the core (derived from the F101 after State Department approval), main control system and design integration. SNECMA took on the LP system,

If you have a billion dollars to get through the first ten years, you can then start making money. CFM International is now enjoying a big income, initially from this CFM56-2 family of engines.

reverser, gearbox, accessory integration and engine installation. All models have a single-stage fan rotating with 3-stage LP compressor driven by a 4-stage turbine, 9-stage HP spool driven by a single-stage air-cooled turbine (about 1,260°C) and annular combustor. The first engine ran at GE on 20 June 1974, and the CFM56-2 was certificated on 8 November 1979.

Large numbers power DC-8-70 series, E-3/KE-3 and E-6 aircraft and, as the USAF F108, KC-135R tankers; French C-135F tankers have the CFM56-2B1. The CFM56-3 has thrust reduced to 20,000 lb from the 22,000/24,000 lb of Dash-2 engines, to suit the 737-300; it has a scaled CF6-80A fan handling airflow reduced from 830 to 656 lb/s. The Dash-5 is an advanced engine in the 25,000-lb class for the A320, with Dash-2 fan, Dash-3 core and digital control.

Charomskii (SOVIET UNION)

Aleksei Dmitriyevich Charomskii was one of the greatest exponents of the diesel aero engine. Beginning work in the 1920s, he was permitted to form a design collective at TsIAM (the national aero-engine institute) and in 1933 ran his first complete engine, the AN-1 (unrelated to the Fiat AN.1 diesel, the designation comes from 'Aviatsionnyi Nyeftyarnoi', 'aviation/oil'. A 2-stroke water-cooled V-12, it probably had cylinders of the same size as the later engines, and was rated at 850 (later 900) hp at 2,000 rpm. From this Charomskii developed the M-30 and M-40, the latter being ahead in timing, running in 1939 and being flight tested in the following year (it is believed, in a TB-3).

Stalin was not interested in strategic bombers but was impressed by the diesel's long-range capabilities, and in October 1940 the M-40 was picked to re-engine the TB-7 (Pe-8). Because of its greater potential power the M-40 was then replaced by the M-30, also first run in 1939 but regarded by Charomskii as nowhere near ready for production. An advanced turbocharged V-12 with cylinders 180 × 200 mm, capacity 62.34 litres, it weighed typically 1,200 kg complete with integral pressure-glycol cooling system, and was rated at 1,400 hp. The production M-30B, from 1941 designated ACh-30B for its designer, was rated at 1,500 hp but caused the pilot problems with the complex fuel controls. Later engines included the ACh-31 and 32 of 1,550 hp, the ACh-39 of 1,800 hp and 39BF of 1,900 hp.

Chrysler (USA)

During the time of violent expansion in June 1941, this famous car company pressured the US Army into giving it a contract for yet another liquid-cooled pursuit engine. The IV-2220 was essentially two inverted V-8s back-to-back with mid-drive to the front gearbox. The 138.75 cu in 4-valve cylinders were highly blown by 2-stage superchargers with a ventral intercooler, and the first run in December 1942 was encouraging. By this time the profusion of competing engines was obvious, and development was abandoned in June 1944 shortly after the XI-2220-11 (the designation had dropped the letter V) had been cleared for flight at 2,500 hp. It was on Chrysler's initiative that Republic completed the first of two planned XP-47H Thunderbolts, in which the

The first Chrysler IV-2220 display model, made in January 1942 (but, of course, not publicly exhibited). The propeller was one of the first to have forged hollow steel blades.

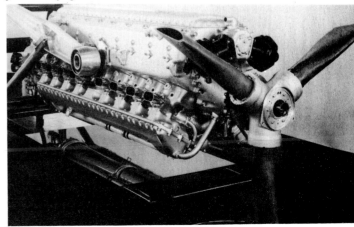

engine was matched with a GE CH-5 turbo, and eventually flew it at Chrysler's expense on 26 July 1945.

Cirrus (UNITED KINGDOM)

Major Frank B. Halford created the first Cirrus engine for Captain de Havilland in late 1924. He was working at Airdisco (the government Aircraft Disposal Company, where among other things were some 30,000 surplus engines). Halford used the cylinders and pistons from one-half of an Airdisco V-8, the air-cooled engine Halford had developed from the wartime Renault. In two months he had designed and built the new 4-in-line, with a 5-bearing crankshaft and deep aluminium crankcase. With cylinders 105 × 130 mm (the Renault was metric), capacity was 4.5 litres, weight 286 lb and maximum power 64 hp at 1,800 rpm, with 68 hp available for take-off at 2,000 rpm. De Havilland flew the engine in the prototype DH.60 Moth on 22 February 1925. Later that year Halford produced the Cirrus II with a bore of 110 mm, forged light-alloy con-rods and bronze valve seats; this gave 85 hp for take-off, followed in 1926 by the Cirrus III with 114 × 140 mm cylinders, rated at 95 hp. Taken over by the Hermes Engine Company, this was marketed with refinements as the Cirrus Hermes I, rated at 105 hp. The 115-hp Hermes II was the first inverted model, all later Cirrus having this layout. The Hermes III and IV were related at 120 to 140 hp depending on sub-type.

In 1934 the business was bought by Blackburn Aircraft at Brough. Later the same year the prototype Cirrus Minor ran at 80 hp, later rated at 90 hp at 2,600 rpm, with cylinders 95 × 127 mm, capacity 3.605 litres. The wartime Minor II had 100 mm bore and gave 100 hp. The Cirrus Major of 1935 had 120 × 140 mm cylinders, 6.3 litres, and was developed from an initial 135 to 158 hp. After World War 2 Blackburn & General Aircraft produced a range of completely new inverted engines, the only one produced in quantity being the Bombardier 4-in-line. This had a magnesium-alloy crankcase and direct injection into cylinders 122 × 140 mm, 6.5 litres, and was rated at 180 hp in fixed-wing installations and up to 200 hp in helicopters.

Top left *Totally different in detail, the IV-2220 installed in the XP-47H exhausted through a turbo (note ducts for piping under fuselage) and had an intercooler in the inlet manifold under the engine. Magnetos are at the front.*

Above left *Frank Halford's Cirrus III was the first of the 4-in-lines to move significantly on from 1916 technology. Rated at 85/90 hp, it went into the 1928 Moths.*

Left *Differing mainly in the crankcase, the Cirrus Hermes I was rated at 105 hp and powered such types as the Avian and Desoutter I.*

Clerget (FRANCE)

Inspired by the Gnome, the Clerget precision-engineering firm produced what it claimed to be an improved rotary engine in 1911 with lower consumption of both fuel and castor-oil lubricant. The main difference was that the mixture was fed from the crankcase via normal external induction pipes to the cylinder heads. The latter were flat, instead of tapered, and housed inlet and exhaust valves both operated by pushrods. Unlike the Le Rhône, the pistons worked directly inside the steel cylinder, and in view of the piston-ring problem (see Gnome) the pistons had special obturator rings, failure of which caused severe cylinder overheating. Clergets were made by the thousand in Britain, as well as France and other countries, and Bentley's first aero job was to seek a solution to this overheat difficulty.

The initial 1911 engine had 7 cylinders 120 × 150 mm, giving capacity of 11.88 litres, the rating being 80 hp at a nominal 1,200 rpm. By 1913 the Clerget 9 was in production; 15.3 litres, rated at 110 hp. In late 1915 the 9B introduced the increased stroke of 160 mm, giving capacity of 16.29 litres, power rising to 130 hp at 1,250 rpm. The final production version was the big 11-cylinder 11EB, nominally of 200 hp. This was priced at £1,663, and even the mass-produced 9B was normally priced on English production at £907.50, considerably more expensive than other rotaries of similar power.

Pierre Clerget kept his company ticking over in Paris during the 1920s, switching to 4-stroke diesels. Some of the first stemmed from the rotaries, though of course they were static radials; the 9A of 1929 gave 100 hp at 1,800 rpm. The 2-row 14F-01, a big 14-cylinder unit developed with government funds, reached 7,665 m in a Potez 25 in 1937. His final fling was the 'Type Transatlantique', a water-cooled H-16 engine with 4 turbochargers designed to give 2,000 hp.

Continental (USA)

In the 1920s Continental Motors was the world's largest maker of engines for use in cars and commercial vehicles built by others. In 1925 it purchased from British Argyll the patent rights to the Burt-McCollum monosleeve valve. It ran, but did not fly, an aero radial using this system in 1927, and subsequently produced a liquid-cooled poppet-valve engine for the Army and an air-cooled sleeve-valve radial for the Navy, neither being flown.

In 1928 the A-70 was run, and by 1929 tested in three different aircraft at 170 hp at 2,000 rpm. This high-quality 7-cylinder radial had bore and stroke both of 4.625 in and a capacity of 544 cu in. It had 2 poppet valves per cylinder, and the engine looked particularly clean because the valve gear was behind the cylinders. From it was derived the W-670 of 1934, with the bore increased to 5.125 in, giving capacity of 668 cu in. This began life rated at 210 hp at 2,000 rpm and during World War 2 was rated at up to 240 hp at 2,200 rpm (250 hp at 2,400 rpm in the equally mass-produced tank-engine version).

A few W-670s are still flying for a living in Ag-aircraft, but Continental's real business started with the emergence in 1931 of possibly the simplest piston aero engine ever built, the A40. This flat-4 (four horizontally-opposed cylinders) had cylinders only 3.1 × 3.8 in, giving capacity of 115 cu in. Single L-type heads were bolted across each pair of cylinders, but the A40 still ran at 2,500 rpm to give 37 hp. It had the important advantage in the Depression of low price, and sold well enough to delay the much better A50 until 1938. The A50 had all the features of bigger aero engines, such as individual cast aluminium-alloy heads screwed and shrunk on to forged-steel barrels, three main bearings and dual ignition. Cylinders were 3.875 × 3.625 in, capacity 171 cu in, and rating 50 hp at the modest speed of 1,900 rpm. In early 1939 the A65 introduced different Marvel or Stromberg carburettors and various refinements, being rated at a useful 65 hp at 2,300 rpm. It was chiefly this engine that resulted in over 8,000 Continental sales by 7 December 1941; after this date a much greater number of military versions were built as the O-170.

After 1945 Continental poured forth refined engines with a standard cylinder 4.0625 × 3.625 in, the C75 and C85 being flat-fours (the number still

Continental's A65 did more than any other engine to establish the flat-4 as almost the standard type for light aircraft.

denoting the horsepower) of 188 cu in and the C115, C125 and C140 being flat-sixes of 282 cu in. In 1947 the E165 and E185 hit the market as the first of the 470 cu in flat-sixes which had the military designation O-470. Over many years the old 'power' designations were gradually replaced by unified civil/military 'capacity' numbers based on cubic inches, prefaced by such letters as 'GTSIO' for 'geared, turbosupercharged, direct-injection, opposed'. Today Teledyne Continental is the world 'No 2' builder of general-aviation engines, its most important family being the O-520 with 6 cylinders 5.25 × 4 in, giving capacity of 520 cu in, with ratings from 285 to 435 hp. For 20 years to 1981 Rolls-Royce Motors of Crewe was a licensee. In 1965 Continental began development of a completely new 'Tiara' range of engines notable for advanced features and in particular for special control of torsional shaft vibration. Despite testing 46 prototypes, and start of production in 1971, the whole programme was eventually terminated, reminding one of the Ford Edsel, Corfam shoes and other classic marketing disasters.

TCM (Teledyne Continental Motors) bounced back at the 1985 Paris airshow with three completely new developments. One was a regenerative gas turbine mainly for non-flying applications. Another was a growing family of piston engines resembling the established models but with liquid (60 per cent glycol) cooling. The third was a series of rotary (Wankel-type) engines, beginning with an amazingly compact 40 hp unit and a twin-rotor engine of 85 hp and geared drive. All may open up new vistas in efficient power.

Less familiar is the story of the engine that was planned as the ultimate in piston aero engines but which was abandoned because it took so long to develop. In 1929 the US Army laboratory at McCook Field carried out research on cylinders giving far

greater power than any others of their size. By 1931 the 'Hyper' cylinder had been finalized with a hemispherical piston top and cylinder head, sodium-cooled valves and 4.625-in bore and 5-in stroke, so that a 12-cylinder engine would have capacity of only 1,008 cu in. The engine was to be an upright V-12 with cooling by Prestone (ethylene glycol) at no less than 149°C, so individual cylinders were used instead of monobloc banks. As the vital cylinder was regarded as already developed, Continental was given the job of producing the complete engine, which the Army fondly thought would take about two years. Harold Morehouse was project manager. In 1934 the Army ordered an increase in cylinder size to 118.8 cu in, or 1,425 cu in for the engine. In 1935 it was decided that future engines would be buried in the wings, for low drag, and so the V-1430 was redesigned as a flat opposed unit, the O-1430.

The first O-1430 was tested at 1,000 hp in 1938; but it could never fit inside a fighter wing, and Colonel Lindbergh's report on the Bf 109, which he found most impressive, resulted in the O-1430 being replaced by the IV-1430, an inverted V-12. It was thought in 1939 that this would soon be ready for use with a turbosupercharger giving 1,600 hp at 3,000 rpm at 25,000 ft, and not only were the Lockheed XP-49 and McDonnell XP-67 designed around it but a giant new company-owned plant at Muskegon, Michigan, was equipped for IV-1430 production. Eventually it was realized the war would be over before production engines could appear, and the IV-1430 was reluctantly terminated—though not before a type-test in July 1944 gave 2,100 hp at 3,400 rpm at 87.8 in manifold pressure!

Cosmos (UNITED KINGDOM)

Cosmos Engineering was the name given to Bristol's

Far left *A half-century of refinement have now led to this 310-hp Teledyne Continental TSIO-520BE, for a Piper Malibu. Each bank of 3 cylinders exhausts through its own turbocharger, and air from the top intake goes through two aftercoolers.*

Left *New in 1985, the IOL-200 is one of Teledyne Continental's new liquid-cooled range. All will later have solid-state ignition and electronically controlled fuel injection. Rating of the IOL-200 is 110 hp.*

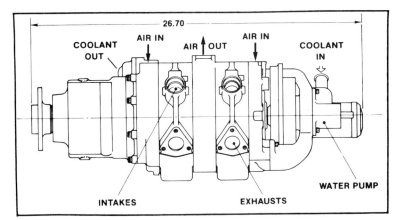

Above right *The dimension in inches shows the compactness of Teledyne Continental's R-36 rotary, to hit the market in 1986-87. The twin-rotor geared engine has a capacity of only 35.8 cu in (588 cc) and is literally smooth running.*

Above right *A rare picture of the Continental 'Hyper' O-1430 flat-12 designed to fit inside the wings of Army aircraft. Project engineer was Harold Morehouse who in 1937 went via Erco to Lycoming, Continental's chief rival.*

Right *When it was realized that the in-wing idea was a non-starter the O-1430 became this inverted V-12, the IV-1430. Output rotation could be reversed by a simple adjustment from outside the drive gearbox. Performance was fantastic, but too late.*

Brazil Straker works at Fishponds after it had been taken over by the giant Cosmos industrial empire in 1918. Brazil Straker's chief engineer, Roy Fedden, had, with L.F.G. Butler, designed two very good radials during the war. First came the Mercury, with 14 cylinders 4.375 × 5.8125 in arranged in helical form and 7 thin conrods working side-by-side on each crankpin. Of 1,223 cu in capacity, it 'ran like a sewing machine' in July 1917 at 300 hp, and was immediately ordered by the Admiralty (but they were over-

ruled by Lord Weir who was captivated by the ABC Dragonfly). In early 1918 came the Jupiter, with 9 big cylinders 5.75 × 7.5 in, 1,753 cu in, aiming at an eventual power of 500 hp. Run on 29 October 1918, the prototype gave about 395 hp for a weight of 662 lb in running order. Fedden put 4 valves in each cylinder for good breathing, but did not dare depart from the poultice type of head, the aluminium casting being added outside the closed top of the steel cylinder. A geared version followed, and direct-drive

Top *The Cosmos (later Bristol) Lucifer was one-third of a Jupiter. The exhaust systems were made laboriously by hand by a skilled one-man subcontractor who became Sir William Lyons, of Jaguar Cars.*

Above *A Cosmos Jupiter of 1919. Each big cylinder had a poultice head with 2 plugs and 4 valves driven by 3 push-rods, everything being exposed. Thanks to Roy Fedden's relentless development, it became world 'No 1' engine after 1925.*

engines powered racers such as the Sopwith Schneider and Bristol Bullet. The Lucifer of 120 hp had 3 cylinders similar to the Jupiter but of only 6.25 in stroke, giving 130 hp at 1,700 rpm, with each firing stroke noticeable. The 1,000-hp Hercules, an 18-cylinder 2-row Jupiter, was about to be built when in January 1920 the Cosmos empire crashed because of a wild financial gamble involving shiploads of household goods for Russia, all of which were of course seized by the Bolsheviks!

Curtiss (USA)

Glenn Curtiss was America's, if not the world's, top racing motorcyclist in the first decade of the century, making his own engines. In 1903 T.S. Baldwin asked him to power a small dirigible, and this flew on 3 August 1904 with an air-cooled V-twin of 60 cu in, rated at 7 hp. In September 1907 he was a founder-member of the Aerial Experiment Association, and provided excellent air-cooled V-8 engines later known as B-8s, with cylinder dimensions reversed, to 3.625 in bore × 3.2 in stroke, 265 cu in, rated at 30-40 hp at 1,800 rpm. In 1908 he switched to cylinders of 5 in bore and stroke in the 50-hp E-4 and 100-hp E-8, one of the latter carrying off top prize (Prix de la Vitesse) at the July 1909 Reims meeting. By this time Curtiss had also built several water-cooled aero engines, retaining the original cylinder construction with a steel barrel and integral head all held to the crankcase by four long tie-bolts and adding brazed water jackets of either copper or non-corrosive Monel alloy.

In 1910 the OX-5 appeared, with bore reduced to 4 in, giving 503 cu in capacity. It retained inclined overhead inlet and exhaust valves arranged in the transverse plane instead of the axial one, the inlet pushrods being inside the tubular exhaust pushrods, and the camshaft driving the magneto and water and oil pumps. The OX-5 suffered from many unnecessary faults, was always unreliable and when made by numerous contractors in vast numbers in World War 1 suffered from appalling quality control. Nominal power was 90 hp at 1,400 rpm, and two important OX-5 powered trainers were the Curtiss JN-4 and DH.6. Smaller numbers were made of V-12 engines of 100, 130 and 160 hp, with cylinders larger than the OX-5.

From early in the century Charles B. Kirkham's machine shop had been much patronized by Curtiss, and by 1912 Kirkham was itching to design an engine himself. Before he saw a Hispano he decided that using a single cast aluminium block for a whole row of cylinders would make an engine rigid as well as light, reduce water leaks and, if a wet liner was used—the water being in contact with the thin steel liner—give improved cooling. The upshot was a joint effort that put the Curtiss K-12 on test in October 1916. This V-12 was one of the truly significant milestones in aero engine development. It had cylinders 4.5 × 6 in, capacity 1,145 cu in, and gave 400 hp at 2,500 rpm. As a by-product, one of the banks of cylinders was used in the low-rated K-6 of 200 hp. The K-12 was in some ways the world's most advanced water-cooled engine of the war period, but the crankshaft needed 7 bearings instead of 4 and the giant monobloc crankcase and the integral cylinder blocks were beyond the state of the casting art, and Curtiss and Kirkham failed to produce a good reduction gear. Moreover the US government decided to standardize on the technically older Liberty; finally, in a row centring on how much K-12 redesign was needed, Kirkham quit, Arthur Nutt taking over.

Curtiss made no attempt to use the C-6 (previously K-6) to compete with the prolific Wright-Hispano, and the few built were mostly derated to 1,700 rpm, giving 150 hp. All effort went into the K-12, which became the C-12 in 1919 (with separate blocks and crankcase but a troublesome reduction gear), the CD-12 (with 7-bearing crankshaft and direct drive) in 1920 and the definitive D-12 which passed its Navy type test in 1922 at 400 hp. In the same year it began a brilliant career not only as an engine for fighters but also for racers, gaining world speed records and Schneider victories, the 1925 Schneider being won with the enlarged V-1400. Fairey attempted to licence-produce the D-12 in England as the Felix (*qv*).

In 1924 Nutt began considering a next-generation military D-12 and in 1926 this entered production as the Conqueror, later also designated V-1570 from its capacity. Bore and stroke were increased to 5.125 × 6.25 in, and a major advance was use of modern open-ended cylinder liners. It was built in many versions with direct or geared drive, often with a supercharger (never used on a production D-12), with ratings of 575 hp in 1927, 600 in 1928 and 650 in 1932. Despite accounting for almost two-thirds of Army expenditure on large engines as late as 1932, Curtiss (merged with Wright in 1939) not only failed to push development but were faced by the Army's rapidly declining interest in liquid cooling. Tests at McCook

Top right *Designed in 1910, the Curtiss OX-5 was still in large-scale production in 1917. It weighed 320 lb, and gave an uncertain 90 hp. This is another Science Museum exhibit.*

Above right *The Curtiss V-1550 of 1926 was an enlarged D-12 which led directly to the V-1570 Conqueror. Note the rear drive to magnetos, camshafts and centrifugal water pump.*

Right *End of the road for Curtiss, the Super Conqueror of 1933 was a massive geared and supercharged member of the V-1570 family which gave up to 800 hp.*

Field in 1931 showed aluminium/steel leakage with high-temperature (149°C) pressurized glycol, and a year later all military funds for Conqueror development were cut off, replaced by long-term support for Continental and Allison.

Curtiss had no success getting airlines to persist with the Conqueror, TAT and Eastern flying water-cooled Condors for only one year. In 1926 Curtiss decided to try to beat the rival Wasp with a totally new aircooled radial, the H-1640 Chieftain. This had 2 rows each with 6 cylinders, the front and rear cylinders being directly behind each other and sharing a common head and overhead camshaft. Diameter was extremely small, and partly because of this it was not very successful. An air-cooled inverted V-12, the Army-funded V-1460, with 4.875 × 6.5 in cylinders, rated at 525 hp at 2,300 rpm, ran well in 1929 but gained no orders. In 1929 Nutt became vice-president (engineering) of the merged Wright Aeronautical division of Curtiss-Wright.

CZ (CZECHOSLOVAKIA)

In 1930 Ceskoslovenska Zbrojovka's Dr Ostroil began development of the ZOD 260-B 9-cylinder radial 2-stroke diesel. Each cylinder had 2 overhead pushrod valves, weight was 598 lb and rating on test in 1933 a reliable 260 hp at 1,560 rpm. Some dozens were made for training and club aircraft.

Daimler-Benz (GERMANY)

This company was formed in 1926 by the merger of Daimler Motoren Gesellschaft, of Stuttgart, the parent company of Mercedes, with Benz. By this time the Allied Control Commission was relaxing its restrictions, and the wartime Mercedes D IIa was built in refined form as well as small numbers of a new 20-hp air-cooled engine derived from the Mercedes F 7502. In 1927 the big F 2 water-cooled V-12 appeared, with cylinders 165 × 210 mm of traditional separate form with welded jackets but with enclosed valve gear. From this Dr Berger's team derived a 750-hp diesel intended for long-distance aircraft. Tested in 1928, this in turn led to the massive LOF 6 airship engine. This was a V-16 more like a marine diesel, with 4-stroke unsupercharged cylin-

ders of F 2 size giving capacity of 54.1 litres. Output was initially 900 hp in 1933, but this rose to 1,200 at 1,600 rpm in the production DB 602 used in *LZ 129 Hindenburg;* in *LZ 130 Graf Zeppelin II* output rose to 1,320 hp at 1,650 rpm, for a weight of 1,976 kg.

Throughout the first half of the 1930s the famous and prosperous firm studied the prospects for a new high-power aero engine, and in 1934 settled on the inverted V-12 layout with glycol cooling. The DB 600 had cylinders of 150 × 160 mm, capacity being 33.93 litres. Each cylinder block was a casting in Silumin (Si-Al alloy) with dry steel liners screwed (and, in pre-war engines, shrunk) into the block. The liner skirt projected and was threaded so that the block could be pulled down rigidly against the internally ribbed dural crankcase, the bottom cover finally being secured by studs and dowel-pins. The supercharged and geared DB 600A weighed 679-687 kg and was rated at 986 hp for take-off and 910 hp at 13,120 ft.

Delivery of prototype engines began in December 1935, the first flight being in the He 118 V2. The He 111 V5 followed in late January 1936 and the Bf 110 V1 flew on 12 May 1936. Luftwaffe expansion resulted in demand far exceeding the capacity of the Stuttgart-Unterturkheim works, and DB production was organized increasingly widely, by 1944 embracing 8 major plants run by the company and 6 run by other organizations. The DB 600 was earmarked for bombers, the He 111B, D and J taking almost all available engines.

The DB 601 was thus eagerly awaited for fighters, notably the Bf 109 and 110, but though the new engine was running in 1935 there was a major bottleneck in production, partly due to poor reliability by both it and the DB 600. The 601 differed in having Bosch direct injection, which in the Battle of Britain embarrassed the RAF by being unaffected by negative-g; other advantages were immunity to choke-tube icing, better behaviour on inferior fuel (87-PN maximum) and claimed lower fuel consumption. The 601 also introduced a fluid clutch drive to the supercharger, which in most DB engines was arranged on the left side and driven by a transverse shaft. Like the long-established Daimler Fluid Flywheel, the oil-coupled drive turned the blower slowly at sea level but slip was progressively reduced almost to zero at rated height. The 601A was rated at 1,100 hp at 3,700 m, take-off power being 1,050 hp. The 10th prototype 601 was prepared as the 601ARJ for the speed-record Me 209 V1, rated at 2,300 hp at 3,500 rpm with methanol boost for 1 minute. The 601E was cleared to run at 2,700 rpm and also had an improved supercharger, while the 601N introduced flat-top pistons to increase compression from 6.9 to 8.2, requiring 96-PN fuel, with rating 1,270 hp at 5 km. This brings the story to mid-1940, where the new

Right *All the Daimler-Benz inverted V-12 production engines were beautifully clean externally. This is a 1,350-hp DB 601E-1, powerplant of the Bf 109F-4.*

Below *The DB 603 was the largest of Germany's mass-produced liquid-cooled engines, with 44.5 litres against the 27 of the Merlin. Almost all the German inverted V-12s had the supercharger running on a transverse shaft.*

projects became so unbelievably prolific, under technical director Dipl-Ing Fritz Nallinger, that they are listed in numerical order:

DB 602, already listed; DB 603, enlarged engine with cylinders 162 × 180 mm, capacity 44.5 litres, rated 1,750 hp at 2,700 rpm and produced from May 1942 (initially in small numbers because Nallinger had not obtained RLM permission and the 603 was officially viewed with disfavour). Subsequently many thousands were built with powers varying up to 2,830 hp at 3,000 rpm (603N); 604, X-24 with cylinders 135 mm square, 46.5 litres, weight 1,080 kg, tested 1940 at 2,660 hp at 3,000 rpm, abandoned 1942; 605, major successor to 601 with blocks redesigned to permit bore of 154 mm, capacity 35.7 litres, selected for Bf 109G early 1942 at 1,475 hp at 2,800 rpm, subsequently 14 production and numerous experimental models including 605AM series with DB 603 supercharger; 606, coupled engine comprising two 601s

side-by-side with inner blocks almost vertical with common reduction gear to single propeller, first flown mid-1937 in He 119, rating 2,700 hp, weight 1,565 kg; DB 607, diesel DB 603, 1,750 hp; DB 609, 16-cylinder DB 603, 61.8 litres, 2,660 hp, 1,400 kg; DB 610, coupled DB 605s as DB 606, 2,950 hp, 1,580 kg, several production models; DB 612, DB 601 with rotary valves, 1,350 hp; DB 613, coupled DB 603Gs, 3,800 hp, 1,993 kg; DB 614, advanced DB 603, 2,000 hp; DB 615, two DB 614 in tandem with contraprop; DB 616, DB 605 development; DB 617, developed DB 607; DB 618, coupled DB 617; DB 619, coupled DB 609, 123.6 litres; DB 620, coupled DB 628; DB 621, DB 605D with 2-stage superchargers; DB 622, DB 603 with 2-stage superchargers followed by turbo-supercharger; DB 623, DB 603G with twin turbo-superchargers; DB 624, DB 603G with totally different series of 2 shaft-drives plus 1 turbo supercharger plus intercooler; DB 625, DB 605D with

Above *Pushing one of 23,000 Bf 109Gs off a Messerschmitt line in 1942. The DB 605A engine was hung from forged Elektron (magnesium alloy) bearers, behind which can be seen the inlet to the supercharger.*

Left *The DB 605D was built in several sub-variants all distinguished by the large supercharger, related to that of the DB 603. Most had water-methanol (MW50) injection and gave 2,000 hp at 2,800 rpm.*

turbo; DB 626, DB 603G with twin turbos and intercooler; DB 627, DB 603G with 2-stage superchargers and aftercooler—survived until March 1944; DB 628, DB 605A with large supercharger added around reduction gear with intercooler before original supercharger entry; DB 629, DB 609 versions with 2-stage superchargers followed by turbo; DB 630, base engine for new series, W-36 configuration with cylinders 142 × 155 mm, 89 litres, 4,100 hp; DB 631, DB 603G with 3-stage superchargers; HZ Anlage, power installation comprising two DB 603S or T on wings supercharged from giant Roots blower in fuselage driven by DB 605T, rated height 13.8 km.

With such a programme brewing it is small wonder that in October 1938 Nallinger refused to enter the field of gas turbines. In 1939, however, he decided he might be at a competitive disadvantage with no such

experience, and he reluctantly agreed to get started. The company rejected Mauch's government scheme that it should take over Heinkel's programme, and instead agreed to develop a very advanced ducted fan, the 109-007. Work was started in September 1939 under Prof Karl Leist, who had joined in January after heading turbosuperchargers at DVL. The 007 was an engine that looks impressive even today. It had two contra-rotating spools, the inner spool having 9 stages of compressor blading and the outer having 8 interleaved stages of compressor blading internally and 3 stages of fan blading externally, the fan being in a full-length duct. There were 4 tubular combustion chambers whose gas was fed to 70 per cent of the periphery of the turbine, the other 30 per cent being cooled by air bled from the fan duct. Design thrust was 1,400 kg at 900 km/h at sea level.

The prototype ran in June 1943 and soon gave 'a

static thrust corresponding to 610 kg at 900 km/h at 7,200 m'. It was cancelled in autumn 1943 because of its long-term nature; in its place DB was told to develop urgently the 109-021 turboprop, using the gas generator of the Heinkel-Hirth 109-011 (*qv*). Intended for a long-range Ar 234, it was to give 3,300 hp at 900 km/h (height unstated).

Dassault (FRANCE)

In 1953 this aircraft company obtained a licence for the Armstrong Siddeley Viper turbojet in its 'long-life' version, rated at 1,640 lb, and developed handed versions for mounting on the wingtips of the SO.9000 Trident, flown on these engines in March 1955. Different installations were used in the twin-Viper MD.550 Mirage I, for which Dassault later developed the afterburning Viper MD.30R. By 1957 the company's own R.7 was on test, this being a Viper scaled up to 25 kg/s airflow and rated at 1,450 kg at 11,800 rpm (compared with 13,400 rpm for the Viper). The afterburning R.7R did not run.

De Havilland (UNITED KINGDOM)

In late 1926 Halford (*qv*) and de Havilland agreed that the Cirrus (derived from the old Renault) should be followed by a 'clean sheet of paper' engine, and the result was the Gipsy air-cooled 4-in-line, run before the end of June 1927. Halford continued to use metric units, cylinders being 114 × 128 mm, and capacity 5.23 litres. The prototype was, in fact, built as a special racing engine tuned to 135 hp at 2,650 rpm on 80-PN fuel, fitted to the DH.71 Tiger Moth monoplane. In 1928 series production began of the Gipsy I, with take-off rating of 98 hp at 2,100 rpm. In the first nine months of 1929 an engine taken at random was sealed and flown in a Gipsy Moth for 600 hours; when stripped, the bill for repairs came to £7. In 1929 the Gipsy II had the longer stroke of 140 mm, giving take-off power of 120 hp at 2,300 rpm. The DH Ghost was a V-8 of 200 hp using Gipsy I cylinders. The Gipsy III was the first inverted model, as were all its successors; it was essentially a Gipsy II, but of course with dry-sump crankcase. In 1931 bore was increased to 118 mm without increasing weight (305

Top right *The first Gipsy I was a 135-hp racing engine handmade to Halford's design by Stag Lane fitters Weedon and Mitchell. The production engine was derated to 85 hp.*

Above right *The Gipsy II appeared in spring 1930, and not only gave up to 120 hp but introduced enclosed valve gear.*

Right *Apart from the redesigned crankcase the Gipsy III of 1931 was virtually a Gipsy II turned upside-down. It was a great success and the IIIA went into production as the 130-hp Gipsy Major.*

Above *Flat discs on the spinners identify the French Ratier propellers on Comet* Grosvenor House, *winner of the 1934 MacRobertson race. They replaced the pilot-controlled Hamiltons fitted for initial test flying. The DH Gipsy Six R engines were also new.*

Left *Biggest de Havilland piston engine, the Gipsy Twelve (RAF Gipsy King) was made only in small numbers (about 95). Halford was distressed that it weighed over 1,000 lb.*

Below left *Last of the de Havilland piston engines, the Queen 70 series began as shown before the war, went into production at 345 hp in 1945 and having been a Bristol Siddeley product finished with Rolls-Royce, at 400 hp. A typical weight was 690 lb.*

lb) and the result was the Gipsy Major I, which at 14,615 engines accounted for just over half the total for all Gipsy engines.

In 1932 a 200-hp engine was needed for the DH.86 and the answer selected was a 6-in-line, the Gipsy Six. This always suffered from torsional vibration, and after its first run—when it vibrated so much it became a blur—had the odd firing order of 1-2-4-6-5-3! The Six was type-tested just in time to fly the first DH.86 on 14 January 1934. Nine months later the MacRobertson race to Melbourne was won by a DH.88 Comet powered by two of the special Six R racing engines with 6.5 compression and high-lift valves, rated at 230 hp at 2,400 rpm. Regular production of the Six II began in 1935, with DH-Hamilton propeller and a rating of 205 hp at 2,400 rpm on 77-PN leaded fuel. Small numbers were made of the Gipsy Minor, with 102 × 115 mm cylinders, capacity 3.759 litres, and rated at 75 hp in 1931 and 90 hp at 2,600 rpm in 1938. The Minor tooling was shipped to Australia on the outbreak of war, but no production took place there. Curiously, DH Australia redesigned their Gipsy Majors for Tigers to have Imperial measures.

An oddball was the Gipsy Twelve, or Gipsy King, an inverted V-12 with regular 118 × 140 mm cylinders, capacity 18.37 litres. It was cleared to run at 2,600 rpm, giving 525 hp, and in the DH.91 and 93 was beautifully cowled with cooling air ducted in the reverse direction from leading-edge inlets. Halford designed the Twelve during his 1935-37 stay in an office in Golden Square, London. Still a freelance, he then agreed to head a design team of 52 at the de Havilland engine works at Stag Lane, where in 1938 he planned a complete new range of Gipsy engines with cylinders 120 × 150 mm, the Major 30 and supercharged 50 of 6.78 litres and the Gipsy Queen 30, supercharged Queen 50 and supercharged and geared Queen 70. Production had to wait until the end of the war, when these new engines, with much greater fin area, and new light-alloy heads and pistons giving 6.5 compression, went into production in fair numbers. The two Majors were rated at 160 and 207 hp, and the three Queens at 250, 295 and up to 400 hp. The last Gipsy of all was the 200-hp Major 200 and its methanol-boosted helicopter version, the Major 215, rated at 215 hp or 222 hp at altitude with turbocharger. Total Gipsy deliveries amounted to 27,654.

Gas turbines

In January 1941 Sir Henry Tizard invited de Havilland to design a jet fighter and the engine to go with it. No specification was issued, but Halford had access to Whittle's work. The decision was taken to use a single engine with the then very high thrust of 3,000 lb, fed from wing-root inlets to a single-sided centrifugal compressor. Another new feature was the use of straight-through combustion chambers, and because of the lack of high-power test facilities there were to be 16 of them. The Halford H.1 was on paper in April 1941, drawings went to the shops at Stonegrove and Stag Lane in August, and the first engine ran in April 1942, reaching 3,010 lb thrust two months later. No airframe existed for it, but eventually two were cleared for flight at 2,000 lb in a Gloster F.9/40 *(DG206)* and flown on 5 March 1943; this was the first flight of a British jet aircraft since the E.28/39 two years previously.

Named Goblin, the engine was flown in the prototype DH.100 at 2,300 lb on 26 September 1943. It was type-tested at 2,700 lb in January 1945 and, as the Goblin II with a new combustion chamber, at the design thrust of 3,000 lb in July 1945. In early 1943

the US Army and Navy were both anxious to have the H.1 built in the USA, and the choice fell on Allis-Chalmers. When Lockheed wrecked the first Goblin in the prototype XP-80 Shooting Star, de Havilland generously removed the engine from the second DH.100 Vampire (the only one available) and sent it by air. It powered the US fighter on its first flight on 9 January 1944. A month later the de Havilland Engine Company was formed, it having previously been a mere division, and Halford at last came on the payroll as chairman. He promptly planned the product line for 1944-45: the H.1 Goblin; an enlarged centrifugal turbojet, the H.2 Ghost; the H.3 centrifugal turboprop of 500 hp; the giant H.4 Gyron axial turbojet for eventual supersonic flight; the H.5, a developed and uprated H.2; the H.6 Gyron Junior; and the H.7 gas producer for helicopter tip drive.

The H.2 Ghost was designed to an airflow of 88.5 lb/s, compared with 63 for the Goblin. It had only 10 chambers, each fed by a pair of tangential elbows from the diffuser. The initial application was the company's own Comet airliner, and this called for a plain circular front inlet, as well as a large bleed manifold for cabin pressurization, an air supply never before attempted on a passenger aircraft. The engine first ran on 2 September 1945, was cleared for flight at 4,000 lb in the outer positions of Lancastrian *VM703*, and later set a height record in a special Vampire. In the latter the Ghost had to have twin diagonal inlets, and in this form it was built in large numbers for most versions of the Venom. Svenska Flygmotor built various models as the RM2, the final RM2B having an afterburner and lateral clamshell nozzle. The H.3 was bench-run only.

The H.4 Gyron was designed to the modest pr of 6, with a 7-stage compressor, most of the compression

The de Havilland Goblin was the first gas turbine to pass a British Type Test, and by mid-1945 had completed a 500-hour test without change of 'any main component'. This picture dates from that time.

Left Bigger and heavier than the engine of Concorde, the de Havilland PS.26-6 Gyron with afterburner was tested in 1955 at 23,900 lb thrust at 6,400 rpm, and the PS.26-3 was to be rated at 27,000 lb. By 1957, when the Ministry halted further work, just over 28,000 lb had been recorded with a simulated P.1121 inlet system.

Left The first-generation de Havilland (Bristol Siddeley) Gyron Junior, the PS.43, was a neat 7,100 lb engine for the Buccaneer, with provision for an enormous boundary-layer control bleed flow.

Left The de Havilland (later Bristol Siddeley) Gyron Junior PS.50 (DGJ.10) was very different from the original Junior. Engine of the all-steel Bristol 188, it is shown with its variable inlet at take-off setting and the 2,000K afterburner detached. The 188 did not carry enough fuel to reach the design Mach number of 2.5.

Left When Westland turned the S-58 helicopter into the Wessex they fitted the Napier Gazelle. Later most Wessex had the Bristol Siddeley Coupled Gnome, seen here, which provided a power reserve of 1,000 hp!

being by ram from Mach numbers around 2. Airflow was 320 lb/s, there were 16 Duplex burners in the annular chamber, the turbine had 2 stages and it was planned to add an afterburner. The Gyron first ran on 5 January 1953. It passed a type-test at 15,000 lb in the same year, was flown in a Sperrin on 7 July 1955 at ratings up to 20,000 lb and was picked by Camm at 27,000 lb with afterburner (PS.26-3 version) for the Hawker P.1121 built at company expense, and finally abandoned after the government minister had insisted the RAF would never need any more fighters. The H.5 was not built, but the Gyron Junior was picked to power the Blackburn NA.39 low-level bomber which became the Buccaneer.

First run in August 1955, the Junior was a 0.45 scale of the Gyron, with 13 spill-type burners and cast blades in the first-stage turbine operating at up to 1,200°C. Take-off rating was 7,100 lb in the PS.43 version produced as the Gyron Junior 101 for the Buccaneer, with a large air-bleed flow for aircraft BLC (boundary-layer control) purposes. The supersonic afterburning PS.50 for the steel Bristol 188 was rated at 14,000 lb.

The H.7 study was the basis of the Napier Oryx, but the DH Engine Company had a very important additional business based on the technology of HTP (high-test peroxide). The first HTP unit was the Sprite ATO (assisted take-off) rocket for the Comet I. Next came the much larger jettisonable ATO rocket pod attached under the wings of the Valiant, the engine being the Super Sprite using HTP/kerosene. Finally the Spectre was a big rocket engine of 8,000 or 10,000 lb thrust, using HTP/kerosene, and produced as an ATO pack for the Victor bomber and as a fully controllable powerplant for the SR.53 and 177 mixed-power interceptors. An advantage of rocket power is that output increases as the aircraft climbs, the reverse of normal, and this was an appreciated feature of a wide range of HTP devices produced for advanced combat aircraft including starters and APUs (auxiliary power units). The company took a licence for the General Electric T58 and ran the first de Havilland Gnome on 5 June 1959, altered to have a Lucas fuel system with Hawker Siddeley Dynamics computer. The licence for GE's larger T64 was not put to use.

Dobrotvorskii (SOVIET UNION)

Aleksei Mikhailovich Dobrotvorskii was permitted in about 1943 to develop a powerful engine using 4 Klimov VK-103 cylinder blocks in X-formation on a common crankcase, the two crankshafts being geared to a single large 4-blade propeller. Designated MB-100, it gave 3,200 hp on test in January 1945. It was installed in a Yer-2 (after that KB [construction bureau] was taken over by

Sukhoi), in a neat 1.95-m diameter cowl, but there is no record of flight. This designer was probably also responsible for the MN-102 engine intended for the Myasishchyev DVB-102DM, left incomplete in 1944; no engine details are known.

Dobrynin (SOVIET UNION)

Little is known about the KB (construction bureau) of Vladimir Alekseyevich Dobrynin. In 1946 he ran the 2,200-hp VD-251 liquid-cooled V-12. It may have been related to the VD-4 under development in 1944-51. An advanced piston engine, the VD-4 had 4 banks each of 6 air-cooled cylinders in 90° X-form, with overall diameter 1.4 m and length 2.5 m. The Tu-85 was powered by 4 turbocharged VD-4K engines each rated at 4,300 hp for take-off, with 3,800 hp max continuous rating. This was qualified in 1951, but was abandoned. By 1955 the KB had passed to Koliesov.

ENMA Elizalde (SPAIN)

Elizalde SA, a major automotive firm, produced licensed Lorraine aero engines from 1925, but in 1929 began considering a range of locally designed engines. First came several variants of the Dragon 5-cylinder radial, with the larger Super Dragons. Immediately after the civil war it re-equipped the Barcelona factory and began a series of Tigre inverted air-cooled in-line engines with cylinders 120 × 140 mm. The 4-cylinder Tigre IVA and IVB, respectively with compression ratio 6 and 6.5 and rated at 125 and 150 hp, remained staple products for many years, but the planned 6, 8 and 12-cylinder models never materialized. In 1944 came the Sirio S-VII 7-cylinder radial with cylinders 150 × 145 mm, initially rated at 450 hp, rising to 500 hp in the 1950s.

The old firm was taken over in 1952 by Enmasa or ENMA, Empresa Nacional de Motores de Aviación SA. Work began on the Beta radial with 9 cylinders 155.5 × 174.6 mm, rated at 775 hp. In 1954 work began on the 275-hp Alción radial for horizontal and vertical installation, with 7 cylinders 110-mm square, followed a year later by the 90-hp Flecha flat-4 with cylinders 105 × 100 mm which ran in 1957. This was the high point. ENMA then began importing Turbo-

méca engines, and most of the piston engines faded
from the scene. Marboré turbojets were made under
licence, but for many years Spain has had no aero-
engine manufacturing industry.

ENV (UNITED KINGDOM/FRANCE)

The ENV, made in two sizes, was one of the best
engines of 1909. It was designed in England but built
at Courbevoie in France, and the name came from the
fact it was '*en V*', French for its configuration of a V-8.
The water-cooled cylinders were 85 × 90 mm in the
small 40-hp size, which sold for £350 in 1910, and 105
× 110 mm in the 60-hp engine, priced at £450. Its
conservative and sound design gave fair reliability,
features including electrodeposited copper water
jackets on cast-iron cylinders, two valves per cylinder
driven from a camshaft above the crankcase (sliding
axially to vary ignition timing and valve lift), a Bosch
magneto and coil and advanced forced lubrication.

F

Fairchild (USA)

After World War 2 this aircraft firm formed a guided
missiles division, and also set up Fairchild-NEPA
(Nuclear Energy for the Propulsion of Aircraft) at
Oak Ridge, to undertake classified research. The
missiles needed a small turbojet, and Ranger Engines
division (*qv*) designed the J44, a simple engine of
24.3-in diameter in a monocoque casing serving
structural and aerodynamic roles and often eliminating the need for a cowling. The diagonal (axial/
centrifugal) compressor handled 25 lb/s at 15,780
rpm, thrust being 1,000 lb and weight 370 lb. Later a
variant was used in booster pods for cargo aircraft.
The 2,000-lb XJ83 did not go into production.

Fairey (UNITED KINGDOM)

Fairey Aviation ran up against a political stone wall in
its several attempts to build aero engines. The first
occasion was 1924 when Richard Fairey was
astonished at the technical advance of the Curtiss D-
12 and Reed metal propeller. He bought licences for
both, and planned a British D-12 as the Fairey Felix.
The D-12 powered the Fox bomber which was so far
ahead of official ideas that it was a grave embarrass-
ment. Watching it fly in October 1925 Sir Hugh

*Forsyth's Fairey P.24 was simpler and much less troublesome than
the Sabre, and also had the advantage that each 1,000-hp half, with
associated propeller, could be shut down. It had carburettors, twin
superchargers and poppet valves.*

Trenchard agreed to order a squadron of Foxes, but
he used the occasion as a stick with which to beat
Napier and Rolls. The latter responded, and the
Foxes of No 12 Squadron were later re-engined with
Kestrels.

Far from giving up, in 1931 Fairey hired Captain
A.G. Forsyth as chief engine designer. Graham
Forsyth had wide experience, and initially produced
the P.12 Prince, a water-cooled V-12 of 1,559 cu in
with a monobloc casting comprising both cylinder
blocks and the upper crankcase. In complete secrecy,
with Fairey's money, three prototypes were built and
tested by 1933 at up to 710 hp or, in the case of the
highly blown Super Prince, 835 hp. In 1934 a P.12
was flown in a Fox II, but Air Ministry policy was that

there was no room for another engine company which 'could never acquire the strength needed to compete with Rolls-Royce and Bristol'.

In January 1935 Fairey studied the P.16 Prince with 8-cylinder blocks and a rating of 900 hp at 2,500 rpm at 12,000 ft for 1,150 lb weight. Instead Forsyth went ahead in October 1935 with the totally new P.24, aimed at carrier-based aircraft. Twin-engine reliability was to be gained (for the first time in any engine) by having two halves each comprising a vertically opposed 12-cylinder unit with a side supercharger, with pressure-glycol cooling. Each crankshaft was geared to its own coaxial propeller, of Fairey constant-speed type. Each half-engine was tested throughout 1938 (the testbed could not handle the 2,200 total horsepower), and on 30 June 1939 the P.24 flew in Battle *K9370*. With a potential of 3,000 hp, the P.24 was considered for the Hawker Tornado and then the P-47 Thunderbolt, the Battle flying some 250 hours at Wright Field in 1942, but wartime pressures eventually forced termination of what was a very promising engine.

Farman (FRANCE)

Between the wars France built so many types of aircraft and engine, most of them in ones or twos, that they defeat the efforts of the chronicler who has to contend with a limited size of book. Farman is an example on both counts. Avions Henri, Maurice et Dick Farman (who were originally British) had plenty of engineering talent in their vast works at Billancourt,

and are credited with two of the chief inventions of the inter-war period: the bevel epicyclic reduction gear, widely licensed to others, and the 2-speed supercharger with clutch and gear-change.

Farman began building aero engines in 1915, and from the start concentrated on finely engineered high-power units running at high rpm, most of them water-cooled and with every conceivable geometric configuration and cylinder size. The only engine made in quantity was the 12WE, with 3 monobloc banks of 4 water-cooled cylinders 130 × 160 mm, capacity 25.48 litres, rated at 500 hp at 2,150 rpm—very like a Lion but heavier at 470 kg and fitted with pushrods instead of overhead camshafts. This flew in an F.60 Goliath in October 1922, powered the Super Goliaths and various other Farman bombers, and kept going for 38 hours in 1924 to set a distance record in the giant single-engined F.62.

Other engines included the neat 8VI inverted V-8 (with added triple Rateau-Farman superchargers for a height record); the derived 400-hp racing version; another racing engine, the 470-hp 12Brs cleared to 4,020 rpm; the 12WI inverted V-12; and two remarkable 18-cylinder engines, the 18 WI inverted engine with 3 banks of 110 × 125-mm cylinders, and the 18T with 3 banks of 120-mm square cylinders arranged in T formation. Both the 18-cylinder engines ran at 3,400 rpm.

Fiat (ITALY)

This great company built its first aero engine in 1908. It drew on racing-car experience with the SA 8/75, an air-cooled V-8 rated at 50 hp with pushrods driving 2 overhead valves per cylinder. Many fresh designs followed, including the water-cooled S55 V-8 of 1912, but the preferred formula from 1913 was the upright water-cooled in-line with large cylinders. By far the most important of the early engines was the A 12, a hefty 6-in-line with bore 160 mm and stroke 180 mm, giving capacity of 21.71 litres. On the left at the midlength was the carburettor group which fed via different geometries of manifold in different versions, while at the back was the vertical shaft driving the overhead camshaft at the top and centrifugal water pump at the bottom. No fewer than 13,260 of these engines were delivered in 1916-19, with ratings from 248 to 300 hp. In 1917-19 Fiat also delivered 500 A 14s, at the time the world's most powerful production engine. A V-12 with cylinders 170 × 210 mm, its capacity was 57.2 litres, weight 791 kg and rated output 725 hp.

Only prototypes were built of the A 15 of 1923, a 430-hp engine with each 6-cylinder block enclosed in a watertight cooling box, and with geared drive from a 2,415-rpm crankshaft. From this was derived the monobloc A 20 of 1925, rated at the same power

The Farman 18WI of 1929 had 3 cylinder blocks cast in Alpax silicon-aluminium alloy. Modest cylinder size permitted a crankshaft speed of 3,400 rpm, at which power was 730 hp. Dry weight with 0.407 Farman reduction gear and Rateau supercharger was 930 lb.

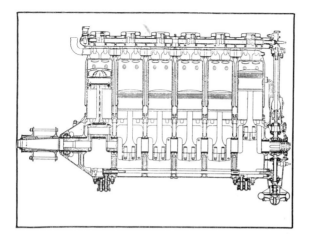

though with capacity reduced from 20.3 to 18.7 litres, and the A 22 of 1926 whose cylinders measured 135 × 160 mm, giving capacity of 27.48 litres and with output typically 750 hp. Several hundred A 22s were built, and they not only gained 13 distance/endurance records but also powered Balbo's transatlantic flying-boat formations. The 700-720-hp A 24 was similar. A much bigger engine was the A 25, with cylinders 170 × 200 mm (54.48 litres), rated at 1,000 hp. Last of the major V-12s was the A 30 RA family dating from 1930, typically rated at 600 hp; these powered Fiat's own biplane fighters, 2,679 engines being built. Inspired by the Gipsy, small numbers were made of the 135-hp A 60 series inverted air-cooled 4-in-line in 1931-34.

In addition, Schneider racing engines were created, the basis being the AS 2 for the 1926 and 1927 races, refined into the AS 5 for the 1929 race. Both were V 12s with 138 × 140 mm cylinders, 25.13 litres, the AS 5 being rated at 1,050 hp at 3,300 rpm. For the 1931 race two AS 5s were bolted together, with a vast carburettor group at the rear feeding into a giant supercharger delivering mixture along a large pipe along the top between the banks. The front engine drove the rear two-blade propeller, and the rear engine gearbox drove a shaft passing between the banks of the front engine to drive the contra-rotating front propeller. Weighing 932 kg this lengthy engine still had less capacity than the A 14, at 50.26 litres, but at 3,200 rpm it put out 3,100 hp to gain the world piston-engined seaplane speed record at 709.2 km/h (440.6 mph) which still stands. Sadly for Italy it did this three years too late for the 1931 Schneider Cup race, when it was still suffering massive and sometimes fatal backfires which were eventually cured by Britain's Rod Banks.

In 1930 the 220-hp AN.1 diesel was flown, a simple 6-in-line water-cooled engine weighing 390 kg. By this time Fiat was firmly wedded to the air-cooled radial, with no more water-cooled projects apart from the RA 1050 RC58 Tifone, the German DB 605A

Top left *This longitudinal section shows the classic simplicity of the Fiat A.12bis, which delivered 328 hp from 6 cylinders—amazing for 1917. The secret lay in the almost-marine size of the cylinders: 160 mm bore and 180 mm stroke.*

Above left *The Fiat AS 2 was a racing derivative of the regular A 14 and A 22 V-12s. Ignition leads were enclosed in a metal conduit which can be seen above the water supply manifold. This was one of the first Fiats with enclosed valve gear.*

Left *In the 1920s Fiat fought Isotta-Fraschini for supremacy in powering Schneider racing seaplanes. The ultimate engine was the AS.5, which lost to Isotta in the Macchi M.67 of 1929, but in 1930-31 it was used as half the mighty AS.6, 2 being mounted in tandem.*

built under licence in 1942. First important model was the A 50 of 1928, with seven 100 × 120 mm cylinders and rated at 95-105 hp. The A 53 of 1930 differed in having a bore of 105 mm and speed raised from 1,800 to 2,000 rpm, giving 120 hp. There were several other small radials, but the main business centred on high-power engines. Several of the early 1930s engines owed a little to Gnome-Rhône and Pratt & Whitney, but the first mass-production type was the 14-cylinder A 74 RC38 of 1936, rated at 840 hp. It led to a mass of 14 and 18-cylinder engines, the biggest with 140 × 165 mm cylinders, all made during World War 2: the 770-hp A 74 RC42, 900-hp A 74 RC18, 1,000-hp A 76 RC40, 1,000-hp A 80 RC41, 1,100-hp A 80 RC20, 1,250-hp A 82 RC40 and 1,400-hp A 82 RC42.

After World War 2 Fiat licence-built the de Havilland Ghost turbojet, and in 1955 produced its own Model 4002, a simple centrifugal turbojet with reverse-flow annular chamber, rated at 250 kg thrust at 26,000 rpm. This assisted design of the Model 4700 gas generator of 542 hp, which served as power unit of the Fiat 7002 tip-drive helicopter prototype of 1961.

Franklin (USA)

Franklin automobile company failed in 1935 and the assets were bought by several former engineering staff, chief engineer being Carl T. Doman. By 1938 the new company, Aircooled Motors, had launched its range of light flat-4 and flat-6 aero engines, built to

Mounted for display, this PZL 6V-350B is one of the last of the Franklins. It is a vertical helicopter engine of 235 hp, all Polish F engines having cylinders 4.625 in by 3.5 in.

a very high standard. First was the 4AC-150, meaning 4 air-cooled cylinders of 150 cu in; it was rated at 50 hp. Subsequently the company concentrated on engines from 65 to 175 hp, with most cylinders 4.25 × 3.5 in and as many interchangeable parts as possible (and no geared drives, to reduce cost). In 1944 the 6ACV-405 introduced vertical fan-cooled units leading to massive sales to Bell and Hiller helicopters. The latter mainly used the 335 Vertical, a 335 cu in engine of 178 or 200 hp in which bore was increased to 4.5 in. The 225, with 4 cylinders of this size, was rated at 75 to 100 hp, and the biggest postwar unit was the 425, with six 4.75 × 4 in cylinders, rated at 245 hp.

Under chief engineer Cregan this range continued to fight for a diminishing market, with various horizontal and vertical engines of 235 cu in (not 225) with 4 cylinders or 335 or 350 cu in with 6 cylinders. In 1975 Franklin gave up and sold all rights to the government of Poland, since when PZL-Franklin engines, now called just PZL-F, have gone into production at Rzeszòw. A flat twin, the 2A-120C of 117 cu in and rated at 60 hp, now complements the 235 and 350 cu in range.

G

Galloway (UNITED KINGDOM)

A subsidiary of Beardmore, Galloway Engineering itself took over the BHP engine and produced it in slightly modified form as the Adriatic. It had a compression ratio of 4.96, compared with 4.56 for the original Beardmore and 5.0 for the Puma, was heavier than either at 690 lb, and used a rotary oil pump instead of two gear pumps. More ambitious was the Galloway Atlantic V-12, the prototype of which comprised two BHP cylinder blocks on a common crankcase, running in about October 1917 at 500 hp. A month or two later the decision was taken to redesign it with Puma blocks, and in this form it was selected for immediate production as the most powerful British engine. Some dozens were made for the DH.15 and V/1500. It was planned to continue production as the Siddeley Pacific, but Siddeley-Deasy produced their own double-Puma as the Tiger.

Garrett (USA)

For many years AiResearch Manufacturing claimed

Above *A cutaway of the Garrett TPE 331 single-shaft turboprop showing the tandem centrifugal compressors, folded reverse-flow combustor and 3-stage turbine.*

Left *Volvo Flygmotor engineers overhaul a Garrett TFE 731 in Sweden. The front fan is driven via a reduction gear.*

Left *Garrett's F109 turbofan, selected for the new T-46A trainer, was first delivered to Fairchild in January 1985. Garrett hopes to market various derived engines.*

to have produced 80 per cent of all gas turbines of 60 to 2,500 hp made in the USA and Europe. In 1957 McDonnell took three of the mass-produced GTC85 turbocompressors and used these to power the Model 120 tip-drive helicopter (which in consequence set a record in lifting a payload double its own empty weight). Since then AiResearch has gradually carved out an ever-bigger niche in the general aviation and light military propulsion market. Since 1981 the AiResearch name has been dropped, though Garrett Turbine Engine Company continue to be based at Sky Harbor Airport, Phoenix, Arizona, where 15,000 aircraft propulsion engines have been delivered since 1965.

Throughout the 1950s the larger engine companies studied powerplants for ultra-high-altitude flight burning liquid oxygen and liquid hydrogen, but the company that got nearest to running a complete engine was Garrett AiResearch. As a result of a proposal by R.S. Rae, a British engineer at Summers Gyro, the USAF awarded Garrett a study contract in October 1955 for what became the Rex I, II and III, the first being a geared turboprop and the Rex III a highly supersonic turbojet with thrust *in vacuo* of 4,500 lb. This scheme was intended for an ancestor of the Lockheed Blackbird family.

Design of the Model 331 began as a private venture in 1959. It was planned as the 500-shp TSE331 for helicopters and as the TPE331 turboprop. The former flew a Republic Lark (licensed Alouette II) on 12 October 1961, but did not go into production. In contrast over 9,900 TPE331s have been sold, including a few military T76 versions. The latter have the reduction gear low with the air inlet above, the reverse of the TPE331. All versions have 2-stage titanium centrifugal compressors (airflow 5.8 to over 8.5 lb/s, pr 8 to over 11), annular combustor and 3-stage turbine with blades cast integral with the disc. The TPE331 is unusual in being a single-shaft engine with no separate power turbine. Weights are 335-400 lb and ratings 575 to 1,645 shp. Next-best for sales (well over 5,000) is the TFE 731 geared turbofan. Announced in 1969, all versions so far have a 3-stage LP turbine driving a single-stage titanium fan connected to a planetary ring gear and a 4-stage LP compressor, a single-stage HP turbine driving the single-stage centrifugal HP compressor, and a folded annular combustor. HP turbine gas temperature is typically 1,010°C, fan airflow 113-143 lb/s, pr (overall) about 14-15 and thrust 3,500-4,500 lb. Bypass ratio is 2.6-2.8 in most engines, but the Dash-5 has a bigger fan raising the ratio to 3.48.

Third production engine is the unique ATF3 (funded initially by the USAF as the F104 for Compass Cope RPVs). It has a reversed 3-spool layout: air enters at the front and passes through the single-stage titanium fan, driven by the 3-stage LP turbine; fan airflow 162 lb/s, bypass ratio 2.8. The core flow of 40 lb/s then passes through the 5-stage IP spool, driven by a 2-stage IP turbine. It continues through the single-stage centrifugal HP compressor (overall pr 25 at cruise), driven by a single-stage air-cooled HP turbine. The HP compressor faces aft at the rear of the engine and its eye is reached via a peripheral bypass duct. From the HP compressor the air passes forwards and then through two 180° bends in the annular combustor, and out via the three sets of turbines, being turned through two sets of 90° cascades in 8 large struts in the mid-section to mix with the bypass flow in the full-length cowl. Accessories are on the circular rear cover. Weighing 1,125 lb, the ATF3-6A is rated at 5,440 lb and, despite its extreme unconventional, has found a modest market in the Falcon 200.

In July 1982 Garrett's TFE76 turbofan was selected, as the F109, to power the T-46A trainer. This could hardly be simpler, with a 28-blade fan, tandem 2-stage centrifugal compressors, annular reversed combustor and 2-stage HP and LP turbines. Weight is 400 lb and thrust 1,330 lb. Such is the progress with high-rpm centrifugal compressors, and in component efficiency generally, that the sfc of this simple engine at take-off is 0.392, while that of the complex ATF3 is 0.506! The TSE 109 is a shaft version (see Allison/Garrett).

General Electric (USA)

It's a small world. When Sanford Moss was a PhD student at Cornell University his gas-turbine experiments offended Professor William F. Durand in the room above. In World War 1 Durand headed the newly formed NACA (National Advisory Committee for Aeronautics), and one of his tasks was to push the development of the turbosupercharger. The world leader in this field was Moss, and this gave GE a valuable background of gas-turbine experience when General Arnold urged NACA to form a jet research group in February 1941. The resulting committee was headed by Durand.

The committee charged three existing steam-turbine firms with building aero gas turbines: Allis-Chalmers, a ducted fan; GE, a turboprop; and Westinghouse, a turbojet. GE had already begun work on a gas turbine for Navy PT boats, but scaled it down to suit the available air supply for testing. This happened by chance to be right for the TG-100 turboprop, and detail design began under Army contract on 7 July 1941. The company's steam-turbine divison at Schenectady under Glenn Warren and Alan Howard used a 14-stage compressor in the TG-100, and multiple tubular combustion chambers. The gas generator ran on 15 May 1943, but a run with

General Electric began design of the TG-100 turboprop before they heard of Whittle. This T31, the outcome, weighed 2,180 lb and gave about 75 per cent of the planned 2,200 shp.

gearbox and propeller did not take place until May 1945. Redesignated T31 by the Army/Navy system, it flew in the nose of the Convair XP-81 on 21 December 1945; the XT31-GE-3 was designed for 2,200 shp plus 600 lb residual thrust, but actually gave about 1,650 shp. The T31 powered the Ryan XF2R-1 and a little work was done on the TG-100B, but the TG-110 and twinned TG-120 were never built.

In March 1941, 10 days after writing to the NACA, General Arnold was in England. He was aware of British jet patents, but was surprised to find the Whittle engine about to fly. He quickly made arrangements to have the W.1 brought to the USA, and picked GE to build it under licence. In April he asked the company to send a good engineer to Britain to work with Colonel A.J. Lyon of the Army Air Corps in London. Such a man was already there: D. Roy Shoults, expert on turbochargers. In September contracts were signed with the GE turbocharger group at Lynn for a Whittle W.1 copy designated Type I (not 'one' but 'eye') Supercharger (as a cover), Donald F. 'Truly' Warner heading the team. On 1 October the W.1X (Power Jets' only bench engine) and a complete set of drawings for the W.2B were flown to Bolling Field, arriving at Lynn on the 4th. GE made many changes, added an automatic control system and used forged Hastelloy B for the turbine blades. The first Type I was mounted in its cell, called 'Fort Knox', and started on 18 April 1942 at 11.05 pm. It ran hot, and on a visit in June Whittle suggested adding partitions in the blower casing to separate the flow to each chamber. This led to the I-A, two of which, rated at 1,250 lb, powered the Bell XP-59A on 2 October 1942.

Thus began GE in the field of aero gas turbines,

starting with nothing but experience with turbochargers and steam turbines yet so motivated and well managed that many people consider it the world No 1 aero-engine company. Even in its first perusal of the Power Jets' drawings it identified features that had proved troublesome, and in autumn 1942 work began on the I-14, of 1,400 lb thrust, with new blower casings, rectangular 90° diffuser passages, fewer but larger turbine blades (as in the W.2/500), a redesigned combustor liner and improved materials. The I-14 gave its design thrust soon after first running in February 1943. The I-16 design was begun in January 1943 and first run was in April; the 1,600 lb rating was soon guaranteed, at sfc of 1.24 and a weight of 849 lb. Altogether GE delivered 30 I-As and 241 I-16s to the Army, as well as a few I-16s for the Navy FR-1. Only prototypes were made of the I-14, of the 1,800 lb I-18 run in January 1944 and the 2,000 lb I-20, run in April 1944. This was because GE and the Army recognized that far more power was needed. The P-59A, with I-16s, was reclassified as a trainer, and in early 1943 the Army asked GE to study a turbojet of 4,000 lb rating. GE did better: it quickly produced two, a centrifugal at Lynn and an axial at Schenectady!

Naturally called the I-40, the Lynn engine was created very quickly in the second half of 1943 by a group under Dale Streid, who in 1939 had studied and reported in detail on the prospects for high-speed jet propulsion by gas turbine. Originally 3,000 lb had been the target for this engine, but in June 1943 GE decided to go ahead with both at 4,000 lb, the centrifugal because it could help win the war and the axial because of its greater future potential. Though designed for airflow of some 78 lb/s compared with barely half as much for the I-16, the I-40 used almost the same compressor and turbine, but with changed materials and manufacturing methods. The big change was the straight-through combustion system with 14 tubular chambers of wrapped Nimonic.

Important advances were also made towards casting turbine rotor blades. The I-40 went on test on 9 January 1944, hit 4,200 lb in February, and flew in the XP-80A on 10 June 1944. This slightly enlarged Shooting Star had been specially designed for the I-40 and the latter soon replaced the J36 (Goblin) as the production P-80 engine. Lynn built 'a few dozen' I-40s, and GE's Syracuse plant worked up close to the planned rate and delivered 300 production engines, designated as the J33, by the end of the war. In September 1945 complete responsibility for the J33 was passed to Allison (*qv*).

The challenging axial engine, the TG-180, drew on the TG-100 for compressor and turbine design and on Whittle for the 8 separate tubular chambers, but its size and power posed major problems, and the manpower needed was at the expense of the less-important turboprop. An 11-stage compressor of constant tip diameter was chosen, with aluminium-alloy discs, steel blades, a magnesium-alloy casing (as in the I-40), airflow of 75 lb/s and a pr of 5. The first TG-180 was run on 23 April 1944 (or, according to recent GE publications, 21 April) at a weight of 2,300 lb, pr of 4 and thrust of 3,620 lb. No insuperable problems were met, and the engine flew at about 4,000 lb rating in the prototype XP-84 on 28 February 1946. Thus, rather surprisingly, both GE's 'own design' turbojets made their first flights in single-engined prototype fighters. By 1946 the TG-180 was beginning to come 'out of the wood' and as the J35 it was adopted for numerous new fighters and bombers. By this time GE had built 140 pre-production engines, and production J35s were beginning to flow from the Chevrolet Division of General Motors, but again Allison was selected for mass-production. Complete responsibility for the J35 was handed over to that company in September 1946, Chevrolet withdrawing voluntarily.

Thus soon after the war's end, GE had lost its two important new turbojets to a competitor. This was done to get Allison into gas turbines and provide competition, and because of that company's available capacity to build. GE now got started in earnest, and under Harold D. Kelsey built up a new AGT (Aircraft Gas Turbine) Division with development and production capacity concentrated at Lynn. Kelsey had to contend with extremely small budgets and with many who believed jet engines would bring only problems and financial losses. But at least the company's forecast was that by 1950 the division's business would total $35 million. To get that business GE decided to design a J35-type engine uprated to 5,000 lb and with better fuel economy.

Under Neil Burgess the TG-190 went ahead on 19 March 1946. It had the same frame size as the J35, and thus could easily replace the earlier engine, but the compressor and turbine were new, the former having 12 stages passing 92 lb/s at pr 5, and so was the lubrication system. There was severe difficulty reducing the weight of major components, and the engine came out heavier than the J35, but it had great potential.

The USAAF accepted GE's proposal for the new engine and for production on a scale rivalling Allison, and began to fund development as the J47. The first engine went on test on 21 June 1947. By mid-1948 production J47s were coming off the line at Lynn, but by this time demand for it was so great that a second source had to be found. The choice fell on Lockland, near Cincinnati, Ohio, a wartime Wright engine plant which had been the world's biggest single building (in terms of volume under one roof). Marty Hemsworth inspected the 40 engine test cells and out of them produced 14 much more capable turbojet cells. An engineering staff of 150 moved in, and on 28 February 1949 Lockland reopened as a GE facility. At first the company needed only part of it, but it later was to be many times extended and today is the aero-engine headquarters.

By 1950 the GE aero business was not $35 million but $350 million. Then came the Korean War, and output had to grow almost explosively, Studebaker and Packard being brought in to build an avalanche of J47s. To the original A-series engine of 4,850-lb thrust came the B of 5,000 lb, the C of 5,200 lb or 6,000 lb with water injection, the D with an improved 100 lb/s compressor and afterburner rated at 7,500 lb, and the E in which several hundred mostly minor

A General Electric I-A, an Americanized W.2B, on outdoor test in September 1942. By this time GE had made more Whittle engines than fund-starved Whittle himself.

changes produced a reliable all-weather engine rated at up to 6,000 lb dry and 6,970 lb with water injection. At the latter ratings the Dash-25 engine powered the B-47E and, together with the F-86 and other types, demanded engines in such numbers that in 1953-54 output from GE, Studebaker and Packard was running at 975 per month. No other turbojet has been produced at such a rate, and the total at completion in 1956 reached 36,500. Late models were assembled vertically, then a novel idea, and the afterburning versions, which concluded with the 7,650-lb J47-33, were the first engines to have an electronic control system, which initially caused severe problems.

In 1947 the newly formed USAF asked GE to study a turbojet of unprecedented size and power. The result was the XJ53, which first ran in March 1951. The XJ53-1, for Mach 1.8 bombers, had a 13-stage compressor handling 264 lb/s at pr of 7.7; it passed a 50-h test in March 1953 at 17,950 lb with turbine entry at 871°C. The Dash-3 missile engine had the same 44-in diameter and 6,500 lb weight but passed 300.5 lb/s at 8.5 pr and ran a 50-h test in January 1954 at 21,000 lb. J53 applications faded, but scaled back to J47 frame size the J53 resulted in the J47-21 (later redesignated J73) of 1949 design under Neil Burgess. This went on test in mid-1950 and repeated the J53's variable inlet guide vanes, cannular combustion, 2-stage turbine and titanium, and the 142 lb/s airflow resulted in a rating of 9,200 lb. Most applications failed to materialize, and only 870 J73s were produced, for the F-86H.

Below *Few jet engines can equal the total of 36,500 General Electric J47s. The most numerous sub-family was the E-series, for the B-47E, this J47-27 being an example.*

Bottom *Festooned with instrumentation, a mighty XJ53 goes to the test cell in March 1951. It was simply too much engine for that era.*

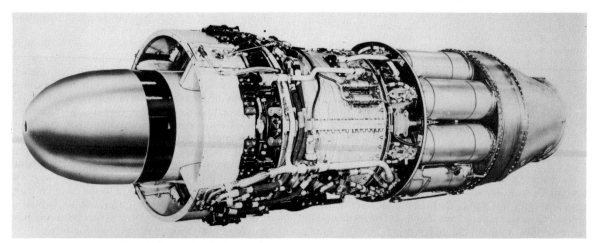

Probably GE's biggest single research programme began with the 1951 contract by the Atomic Energy Commission and USAF for a five-year study on nuclear shielding requirements to help determine if a nuclear-powered aircraft was feasible. By 1954 the study hardened into a team effort with Convair, in competition with P&WA/Lockheed, for propulsion of the WS-125A NPB (nuclear-powered bomber). In June 1955 GE formed a task force to work on a direct-cycle engine in which the reactor replaced the normal combustion chamber, the engine airflow actually passing through it. The project was headed by Roy Shoults; the engine programme was managed by Bruno Bruckmann (see BMW) and another key figure was a fellow-countryman, Gerhard Neumann, destined to be GE's greatest aero engine leader. Neumann ran a modified J47 on nuclear heat, but he eventually picked an unusual arrangement in the X-211 (J87) engine in which two colossal afterburning turbojets were linked on each side of one reactor, half the 300 lb/s airflow of each engine passing through the reactor and the rest bypassed via giant pipes. Several forms of reactor were tested, with cores at up to 1,100°C, and it was the thorny problem of how to service an intensely radioactive engine that led to the X-211 bypassing the problem by keeping the engine hardware outside the core. The X-211 compressor set new records with multiple variable stators and a pr exceeding 20, the overall engine being 41 ft long and having sea-level thrust for each half of up to 27,370 lb, achieved in January 1961. The NPB project was terminated on 31 March 1961.

Long before this time GE knew it had a new engine that was a world-beater. In 1952 the AGT Division was led by C.W. 'Jim' LaPierre, who urged continued technical leadership. What was wanted was a dramatic advance in the next major turbojet, with good fuel economy at Mach 0.9 but also the structural strength and high thrust for Mach 2, combined with reduced weight. The central route to all these objectives was increased pressure ratio, and in 1951 Neumann decided to go for the variable-stator compressor. The idea was that most of the stages of stator blades would be mounted in rotary bearings and, coupled by rings of linkages, be driven to the best angle for the engine operating condition by a fuel-powered hydraulic jack. His group built a VSXE (variable-stator experimental engine, inevitably known as the 'Very Sexy') and in March 1952 completed the layout drawing for the proposed X-24A engine. It looked impressive, but competing teams were set up; these quickly rejected bleed valves, but continued to study the dual-rotor (2-spool) arrangement. In October 1952 a management conference in Indiana under LaPierre took the fateful decision: go ahead with the variable-stator engine.

They may well smile as they hustle the first XJ79 to the test cell in June 1954: almost another 17,000 followed it!

In November the X-24A was accepted by the USAF as the J79, the GE unclassified designation being MX-2118, and the rival Advanced J73 and J77 were dropped. To support the radical new engine the GOL-1590 demonstrator was designed to weigh only 2,935 lb and give afterburning thrust of 13,200 lb—figures beyond any previously attempted. The first test compressor was finished in August 1953 and gave results so high it was thought the instrumentation must be faulty (it was not). The GOL-1590 itself was started at 5.00 am on 16 December 1953. Power was slowly brought up to the maximum; suddenly there was a deafening explosion and the front of the engine virtually disintegrated, the rest screeching to a stop. The cause: a faulty 'dog-bone' link holding the engine to the bed! The repaired GOL-1590 was back on test on 15 January; it ran as predicted. The J79 (MX-2118), whose original project head, Perry Egbert, had been replaced for health reasons by Neil Burgess, followed not far behind and the first went on test at Lockland on 8 June 1954. First flight was in a retractable pod under a B-45 at Schenectady on 20 May 1955. On 8 December 1955 GE chief test pilot Roy Pryor took off from Edwards in an XF4D powered by the new engine.

Since then 16,950 J79s have been built, 3,249 of them assembled by licensees, in a programme which has brought GE $4.5 billion and will require service support into the next century. The engine set 46 world records, and so far has logged 35.9 million hours. Features include a 17-stage compressor with 7 variable-stator rows handling airflow of 166 to

Left *GE made about 17,000 fewer of the immense X-211 nuclear powerplant. The twin J87 turbojets had chemically-fuelled afterburners but used a high-temperature reactor instead of the usual combustion chamber. Problems were awesome.*

Right *A twin-spool high by-pass ratio unit, the TF39 turbofan was first demonstrated in December 1965 and made its initial flight on a modified B-52E on 9 June 1967.*

Left *There are still a handful of CJ-805-23C turbofans flying. This aft-fan engine was created by adding a rear section with double-deck turbine/fan blades, downstream of a regular CJ-805 commercial turbojet (itself a J79). Note the linkage to the variable stators.*

Below left *A J85-21 for the F-5F. Weighing 684 lb, this diminutive afterburning turbojet has an airflow of 53 lb/s and gives 5,000 lb thrust. Installed, it fits downstream of the auxiliary inlet doors seen open near the tail! Note GE overalls.*

Right *The GE4/J5P, engine of the Boeing 2707-300 SST, was the most powerful aircraft turbojet ever built. Features included a 9-stage variable-stator compressor handling 633 lb/s, 2-stage turbine for continuous operation at over 1,100C and an afterburner and nozzle of awesome proportions. Weight was 11,300 lb; thrust hit 69,900, equivalent at Mach 2.7 to 333,000 hp.*

170 lb/s at pr 12.9 to 13.5, a combustor with 10 cans in an annular chamber, a 3-stage turbine, and a long afterburner with profiled petals positioned by 4 rams driven by lube oil. Thrust rose from 14,350 lb in the first version to 17,820 lb in most current J79s. Later the LM-1500 version was developed for surface applications, while in 1956 GE launched both a simplified civil version, the CJ-805, and a novel aft-fan derivative. The CJ-805 eventually entered service in 1960 with the Convair 880, but it was a small programme. The aft fan, masterminded by Peter Kappus, had the advantage that it left the engine upstream unaffected by the large-diameter addition of a free-spinning ring of 'bluckets', so-called because each comprised a turbine blade (known as a 'bucket' in US parlance) carrying a superimposed fan blade handling fresh air in a short duct at the rear of the engine pod. The resulting CJ-805-23 raised thrust to 16,050 lb from the 11,200 lb turbojet, and dramatically improved sfc and noise; but again the marketing proved a disaster, to both GE and Convair, and only 37 fan-engined CV-990 Coronados were sold.

In 1953 GE's AGT division had diversified and reorganized, while the giant Ohio plant had been renamed Evendale, from a newly incorporated village in suburban Cincinnati. Meanwhile, at Lynn, studies of small engines led to a Navy contract for a helicopter turboshaft to weigh 400 lb and deliver 800 hp. GE like to beat the requirement, and in December 1955 the prototype T58 actually weighed 250 lb and delivered 1,050 hp. Subsequent versions reach 1,800 hp. Novel for a small engine in having a 10-stage axial compressor with 4 variable stators (handling 12.4 to 14 lb/s with pr of 8.4), it was ordered in large numbers, licensed to DH (now Rolls-Royce) as the Gnome and to Alfa Romeo and IHI. It also spawned the commercial CT58 and marine/industrial LM100.

In October 1953 the Small Aircraft Engine Department was formed by Jack Parker and Ed Woll. Its product was merely the T58, so in 1954 Woll started small-jet studies and proposed to the USAF engines of 10:1 thrust/weight ratio (2,500/250 were the figures) with pr of 5, 7 or 12. Back came a contract for the J85, with pr 7; it was wanted only for the GAM-72 Quail decoy missile, but under Fred MacFee the J85 soon gained manned applications, most notably in Northrop's T-38 and F-5 family. Today over 13,500 J85s have been delivered, as well as over 2,000 commercial CJ610s and over 1,100 derived CF700 aft-fan engines. The latest J85s are rated at 5,000 lb with afterburner, final models of CJ610 and CF700 respectively giving 3,100 and 4,500 lb.

In 1955 the USAF was funding two rival engines for the extremely fast F-108 interceptor and WS-110 CPB (chemically powered bomber). Neither engine was a GE product, yet by a combination of impressive

A claimant to the title of 'most powerful engine', the first CF6-80C2 ran in May 1982 at 62,000 lb. Airflow is 1,754 lb/s.

technical design and willingness to fund demonstrator hardware, GE came from behind, and in May 1957 received contracts for its J93 engine to power both aircraft, the bomber having become the XB-70. The J93 was GE's biggest challenge to date. A high-pressure variable-stator engine, its first-stage turbine blades set a new high in temperature and introduced the STEM (electrolytic machining) method of drilling holes now used on almost all advanced engines. The J93-3 ran on special high-temperature JP-6 fuel and produced 27,200 lb of thrust in full afterburner for a weight of 4,770 lb. But the definitive J93-5 was designed to burn high-energy 'zip fuel' in its afterburner based on ethyl borane. Over $200 million was spent, mainly on USAF budgets, setting up a zip-fuel industry. The technical problems were desperate, and the hated boron-based fuels were abandoned in August 1959, followed by the F-108 a month later and the B-70 in December 1962.

Back in 1953 what became the T58 had been studied as the basis for turboshaft, turbofan, turboprop and turbojet engines for a wide range of contrasting aircraft. In 1961 Neumann asked MacFee to study the market for an engine between the J85 and J79. This led to an unprecedented application of the building-block concept, resting on a basic core engine, the GE1, developed under Jim Worsham.

The GE1 itself, with a 14-stage variable-stator compressor, gave 5,000 lb thrust but was about half the length or frontal area of a J47; as the J97 it later flew in Compass Cope RPVs. Though little known, it was the starting point for almost all GE engines launched since that time. They include two seemingly totally different engines bigger by far than anything previously seen.

First came the GE4, for the American SST launched by President Kennedy in May 1963. It stemmed from the X279M of 1958, which was a variant of the J93. Under John Pirtle the GE4 quickly took shape as a scaled GE1 with J93 features, with hollow compressor blades, the biggest turbine (with air-cooled blades) and an afterburner tested by blasting two J79s into it. At 69,900 lb it was the most powerful jet engine known, but it died along with the SST on 29 March 1971.

More survivable was the TF39, the world's pioneer HBPR (high bypass ratio) turbofan. Under General Bernard Schriever the USAF had in Project Forecast in 1962 asked for masses of long-range industry data. GE provided truckloads, including input from its tip-drive lift-fan produced for the Ryan XV-5A VTOL. Sifted, the answer was that future transports needed HBPR engines—a fact clearly evident from Whittle's pre-war calculations. The USAF started its project for a gigantic airlift transport, the CXX, and GE's Don Berkey submitted a proposal for a remarkable engine of bypass ratio 8, with a '1½-stage' fan comprising a single fan stage, without inlet vanes, followed by a second stage of smaller blades with 'split work' aerodynamics to supercharge both the fan and core flows. The USAF liked this, but in March 1964 insisted that GE must demonstrate an engine.

The GE1/6 was built and demonstrated with sfc of 0.336 and the USAF accepted this, even though at 15,830 lb it was half the power of the required engine. To win the CXX competition GE had to produce 50 copies of a 90-volume proposal, and three days before the deadline the USAF asked for many extra volumes filled with specific information for evaluation teams. When oral presentations had to be made, in April 1965, GE sent the whole 'First Team' from Parker and Neumann down; it was worth it, and in October 1965 the AGT division won its biggest single contract, $459,055,000. The TF39 posed such challenges as a fan airflow of 1,549 lb/s, a 25-pr compressor and 1,371°C turbine, but today the total flight time on the original 81 aircraft is about 2.92 million engine hours and the TF39-1C engine is back in production for the C-5B.

In 1954 the Small Aircaft Engine department began development of a larger turboshaft/turboprop with better sfc than the T58. The result was the T64, which flew as a turboprop in 1960 in the Caribou and as a turboshaft in the CH-53 in October 1964. Another fixed-wing application is the G222, for which 188 T64s will have been licence-built by Fiat by late 1986.

In 1967 work started on projects that were to lead to five new families of engines, on which most current business is based. Biggest was the giant commercial turbofan. When Boeing/PanAm were seeking an engine for the 747, GE consciously 'walked away', aware of the fact that the CTF39—derived from the existing TF39—was not sufficiently powerful, and fearing the impact on the TF39 of the drain on resources needed to produce a bigger engine. Douglas and Lockheed were busy with twin-engine 'airbus' studies for which Rolls were offering the big RB.207, but the GE Commercial Engine Projects team, formed under Ed Hood, succeeded in persuading the two US planemakers to consider trijet aircraft powered by the CF6 engine rated at 32,000 lb.

The engine went ahead on 11 September 1967 under Britisher Brian H. Rowe, but quickly had to grow, first to 40,000 lb as the CF6-6D for the DC-10-10 and in 1969 to 49,000 lb as the CF6-50 for the DC-10-30. The basic Dash-6 engine has a '1¼-stage' fan, 16-stage HP compressor, 30-nozzle annular combustor, 2-stage 1,330°C HP turbine and 5-stage 871°C LP turbine, weight being 7,896 lb. Several navies and some merchant ships use the derived LM-2500. The Dash-50 has a fan rotating with 3 LP compressor stages with free-floating bypass doors, 14-stage HP compressor handling airflow increased from 194 to 276 lb/s and only 4 stages on the LP turbine. Original fan airflow was about 1,300 lb/s and overall pr 24.3, the latter rising in the Dash-50 series to between 28 and 30.

The first CF6 ran on 21 October 1968 and three of them quietly propelled the first DC-10 on its ceremonial rollout in July 1970. The CF6-50 first ran in September 1970. Right at the start, in March 1968, GE were shattered when the first three trijet customers to announce a choice of engine all picked the RB.211, but gradually GE's ability to offer an uprated engine made it No 1, the choice of Airbus Industrie and even of 747 customers—which the company had not counted on. At the time of writing, total CF6 flight time exceeds 35 million hours. The advanced-technology CF6-80 family are in service in the A310 and 767, and the completely redesigned new-generation CF6-80C2 exceeded 62,000 lb thrust on its first run in May 1982, flew in an A300B4 on 19 August 1984 and is the subject of a technical collaboration agreement with Rolls-Royce.

Second of the 1967 launches was the TF34 turbofan for the US Navy S-3A anti-submarine aircraft, with a single-stage fan (338 lb/s), 14-stage compressor scaled up from the T64, and rated at 9,275 lb thrust for a weight of 1,478 lb and with sfc of 0.363 (better than most CF6s). Developed within a tight fixed budget, the TF34 later powered the A-10A and is now finding civil markets as the CF34. Third 1967 project was the GE12, a 1,500-shp turboshaft demonstrator aimed at a major US Army helicopter competition. The compressor comprised 5 axial stages followed by a centrifugal, and a key factor in GE's win was not only the sustained good performance of the engine in extremely adverse test conditions but also advanced design features to reduce the number of parts—for example by making each axial stage from a single slab of steel, machined to form a 'blisk' (blades plus disk)—and minimize overall costs. The GE12 led to the T700, now flying in large numbers in several kinds of helicopter.

Fourth of the 1967 projects was the GE15, a neat 2-spool turbojet that GE thought might lead to its urgently sought 'next J79'. Derived from the GE1, the GE15 was a low-bypass afterburning engine to be rated at 14,330 lb and sized to replace a modified J97 as the engine of Northrop's P-530 Cobra. GE, recognizing a massive gap between the J85 and P&W's F100, offered to develop the GE15 comprehensively for the Department of Defense at a cost of just $10 million. This led to the YJ101 which powered the YF-17 prototypes of 1974. For the second time (the first was in the F-14/F-15) GE lost to P&W in the lightweight fighter competition, but unexpectedly the Navy finally picked a more powerful derived engine, the F404, for its F/A-18 Hornet. The F404 has a bypass ratio increased to 0.34, an airflow of 142 lb/s and is rated in the F/A-18 at 16,000 lb for a weight of 2,180 lb. From the start it has proved an outstanding engine, and it has since gained other applications, most notably including the Swedish JAS39 in which an uprated version is built by Svenska Flygmotor (*qv*).

Fifth engine launched in 1967 was a derivative of the GE9, one of the GE1 offshoots. A large augmented turbofan, it drew heavily upon the X370 demonstrator created in 1960-61 by John Blanton's team to combine 10:1 thrust/weight ratio with the ability to fly at Mach 3.5, with turbine temperatures explored into the 2,200°C region. The work paid off and in June 1970 the USAF selected the GE engine, which had become the F101, to power the AMSA, later the B-1. Under Worsham the F101 core ran in October 1971, and the complete unaugmented engine in January 1972. From it stemmed the F101-GE-102 which powers the B-1B, an engine in the 30,000-lb class with airflow of some 350 lb/s. Moreover, GE never gave up on big fighter engines, but built a series of F101 DFE (derivative fighter engine) demonstrators which were flown in the F-14 Super Tomcat, F-16/101 and F-16XL. This work also paid off; the 101DFE became the F110, a splendid engine

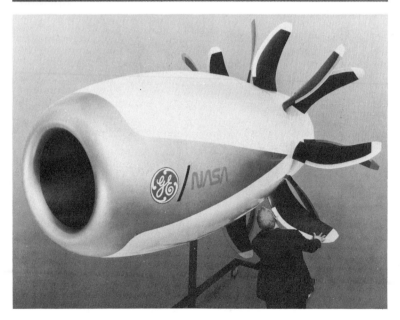

Above *GE's CT7 turboprop powers several new commuter liners, at ratings up to 1,725 shp. The power section is basically a helicopter T700.*

Left *The sheer excellence of the F110 fighter engine has been rewarded by its selection not only for the F-14Ds but also future F-16s and (possibly) some F-15Es. It is in the 28,000 lb thrust class.*

Left *A mock-up of the General Electric GE36 UDF (unducted fan), very much an engine of tomorrow. Boeing expect to fly one in No 3 position on a 727 in 1986.*

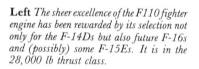

Right *One of the first series of production Gnomes, a 50-hp Omega 7-cylinder (part-sectioned) seen from behind.*

in the 28,000 lb thrust class with airflow of some 260 lb/s. After prolonged competitive evaluation the F110 was ordered in 1984 to power future production F-14s and F-16s, breaking the P&W monopoly and probably putting GE into the position of world 'No 1' engine builder.

Today under Brian H. Rowe the AEBG (Aircraft Engine Business Group) is headquartered at what is now called Neumann Way, Evendale. It has two main operating arms, one (previously Commercial & Military Transport Engines) in Ohio and the other (formerly Military & Small Commercial Engines) at Lynn. Lynn announced in September 1976 the CT7, a civil family derived from the T700 and produced in both turboshaft and turboprop forms in the 1,700-hp class. Also at Lynn is the most recently announced engine, the GE27 demonstrator in the 5,000-hp class expected likewise to mature in both shaft and prop forms. One of the most exciting future prospects is the propfan, which combines propeller efficiency with jet speed. GE has every intention of leading the field in this new technology and is fast developing the GE36 as the first of what it calls UDF (unducted fan) engines. The core is an F404, and without a gearbox it drives contra-rotating propfans at the rear.

GE is one of the two principals of CFM International, and in 1982 signed a collaborative agreement with Rolls-Royce on the CF6-80C2 and RB.535E4.

Glushenkov (SOVIET UNION)

Little is known of this design bureau, announced in 1969 as having produced the TVD-10 turboprop for the Be-30. It was later realized that the same gas generator first ran, probably in 1960, as the GTD-3 for the Ka-25 family of coaxial helicopters. It has a compressor with 7 stages, 6 axial and 1 centrifugal. Even the centrifugal rotor has inserted blades, with airflow of 4.5 kg/s at 28,800 rpm. The annular combustor has two starting units which incorporate auxiliary burners, and there are separate 2-stage compressor and single-stage power turbines. Maximum (except contingency) ratings are 900 hp for the GTD-3 and 990 for the GTD-3BM. The TVD-10B is the derived turboprop, which gives 960 shp but weighs 650 lb compared with 307 lb. It powers the An-28 and is in production in Poland as the PZL-10S, the Rzeszòw factory also licensing the helicopter version for civil use as the PZL-10W.

An uprated derived engine, the 1,450-hp TVD-20, is likely to be built in very large numbers as the engine of the biplane An-3. It will probably also be licensed to Poland.

Gnome (FRANCE)

The Gnome took the Paris-centred aviation world by storm when it was marketed in spring 1909.

Designers Louis and Laurent Seguin had built conventional Gnome petrol engines from 1892, but studied the profusion of rotary engines in which the cylinders revolve round a stationary crankshaft. Aviation examples included Hargrave (1888, compressed air), Burlat (1905) and Adams-Farwell (picked by Berliner for his 1907 helicopter, though really a car engine). They may have designed their Gnome rotary aero engine as early as 1907, but it was revealed in about November 1908. The first version had 5 cylinders of 100 mm bore and stroke, 3.4 litres, weighed 60 kg and is said to have developed 34 hp at 1,300 rpm. According to some accounts it was fitted to Roger Ravaud's hydroplane at Monaco in January 1909 (other reports say the 7-cylinder was used), which, curiously, had contra-rotating propellers; not surprisingly, it failed to fly. The first Gnome to fly thus appears to have been the 7-cylinder of 110 mm bore and 120 mm stroke, 7.98 litres, rated at nominal 50 hp at 1,200 rpm, bought by L. Paulhan and flown in his Voisin on 16 June 1909. In the close-knit Paris *aviateur* fraternity the Gnome spread like wildfire and several more were quickly sold. Henry Farman actually removed his Vivinus during the Reims meeting and with the Gnome won the Grand Prix de Distance on 27 August and the Prix des Passagers on the 28th!

Not only was the Gnome novel in being a radial rotary but in these early pusher applications it was actually mounted behind the propeller! The weight, thrust and vibratory loads were all taken by the tubular crankshaft, bolted to the airframe. Through it

mixture from the carburettor passed to the crankcase, from where it reached the cylinders through valves in the heads of the pistons. These valves were forced shut by the combustion pressure; on the return stroke they were opened by a pair of pivoted counterweights inside the piston. Exhaust gas was expelled straight to atmosphere through a valve in the centre of each head, driven by a cam ring and push rod. Construction was almost entirely machined from solid forged nickel-steel, to a high standard. The master rod had a big end in the form of two large discs, each running on the crankpin with its own ball bearing and carrying holes for the wrist-pins of the secondary or link rods.

The whole engine had to be stressed to withstand up to over 100g because of the rotation, cooling of the cylinders was uneven, and no normal lubrication system would have worked. Cylinder distortion was partly overcome by adding a very flexible obturator ring, made of thin bronze, above the normal piston rings. For lubrication, oil was added to the fuel mixture in quantities of 25 to 35 per cent! The result was that the head of each cylinder spewed fire and unburned oil, so a cowling was essential in tractor installations. As the oil had to be imiscible with petrol a vegetable-based type (usually castor) was used, the aroma being as inseparable from rotaries as kerosene was from early turbojets. Advantages were light weight, typically 75 kg for the 50-hp (which usually gave about 40 hp), automatic damping of propeller

One of the later Gnome rotaries, a BB 18c (double-9) rated at 240 hp. This was one of the monosoupape *family.*

vibration without the need for a flywheel, and adequate air cooling even on the ground. Well-maintained, reliability was fair, but the valves in the pistons caused dangerous backfires through the carburettor, and often complete failure.

By 1910 the basic 7-cylinder had been enlarged to about 70 hp, while the original size had been doubled up to provide a 14-cylinder 2-row unit of 100 nominal hp. The Seguin brothers also studied ways of eliminating the troublesome piston valves, and in 1912 came out with the Gnome *Monosoupape* (ie, single-valve). Holes were simply cut in the cylinder walls uncovered by the piston near the bottom of its stroke, through which very rich mixture could be admitted. The operating cycle was as follows: ignition 20° before top dead centre (TDC), useful power from TDC to 90°, exhaust (head valve) open from 90° right round to TDC (total 270°), induction of pure air (through exhaust valve, still open) from TDC to 135°, the continued downstroke then producing a partial vacuum into which rich mixture was admitted from 20° before BDC (bottom dead centre) to 20° after, followed by compression of the diluted mixture to the ignition point. Most Gnomes, prior to the Mono, had a throttle lever as well as a fine-tuning mixture control, but no control over ignition timing. The Mono, however, had no throttle, and the so-called fine-adjustment lever gave extremely coarse control; accordingly a blip switch on the control column enabled the pilot to cut ignition in or out (often on any selected number of cylinders) at all times on the ground and during the approach.

Thus by 1914 Gnome, and their agents in all major countries, had a large range of rotary engines. The original pattern, often assigned Greek letters by the company such as Lambda (and, for the corresponding 2-row engine, Lambda-Lambda), was marketed chiefly as the 80-hp (actually about 62) with cylinders 124 × 140 mm and priced in Britain at £430, and in small numbers as the 100-hp 10-cylinder and the 160-hp 14-cylinder. During the war the Mono engines were more numerous, the chief models being the 80-hp, the 100-hp with nine 110 × 150 mm cylinders, 12.83 litres (actually giving about 103 hp at 1,200 rpm), the rarer 150-hp with nine 115 × 170 mm cylinders, 15.89 litres, actually giving about 157 hp at 1,250 rpm, and important 18-cylinder engines of 205-240 hp.

The Japanese Shimadzu was a direct copy of the original Gnome series, while the German Oberursel began as such but later copied Le Rhône. Other rotaries, such as the Clerget, Le Rhône and Bentley, are discussed separately.

Gnome-Rhône (FRANCE)

In 1914 the Gnome company bought out its rival to

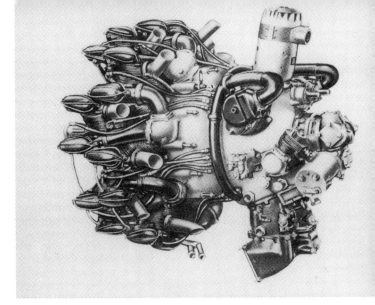

form the Société des Moteurs Gnome et Rhône, though during the war the Le Rhône design team were allowed to get on with their own engines. There appear to have been no Gnome designers other than the Seguins, and in 1921 they were anxiously wondering how to keep producing saleable engines. In October of that year they wisely took a licence for the new Bristol Jupiter. With it came two Bristol resident staff: Norman Rowbotham was appointed chief engineer and general manager and Roger Ninnes became chief designer. They were joined by another Englishman already at the Boulevard Kellermann factory testing the company's motorcycles: Ken Bartlett. All three headed the GR aero team through the 1920s, and all later became directors at Bristol, Bartlett on the aircraft side.

In 1925 GR introduced the Farman reduction gear, and by this time was looking for ways to evade the terms of its licence. In 1927 the company began not only building the Bristol Titan, without paying a royalty, but it began developing it into GR engines, first into the 7-cylinder Titan 7Ksd (almost a copy of the parent firm's Neptune), then in October 1927 into the 9-cylinder Mistral, and finally, in December 1928, into the first Bristol-derived 2-row engine, the 14-cylinder Mistral Major, still with the 146 × 165-mm cylinder but with capacity 38.7 litres (the same as Bristol's later Hercules). In order to achieve reasonable valve gear the Major had only 2 valves per cylinder, all valves being operated by sloping enclosed pushrods from inlet and exhaust cam rings at the front of the crankcase. GR were ahead of Bristol in enclosing the valve gear (on a separate forged head on an open liner) and at least equal in providing greater area of cooling fins per cylinder. The Mistral Major, known as the 14K with suffix letters for sub-variants, began life at a weight of some 660 kg and power 750 hp, and by 1932 was in major production at powers between 800 and 900 hp, which was higher than any other production radial of its era.

Thus, GR managed to sell a lot of engines and even licences, to Italy, Japan and the Soviet Union, among others, and by the 1930s their own engineering staff was large and capable. The 14K was progressively developed until 1953, via the extremely important 14N family of 900-1,100 hp of 1935-50, the 1,500-1,600-hp 14R of 1939-50 and the post-war 14U of 2,200 hp. In 1945 the company vanished into the nationalized group SNECMA, and the final member of the family was the 1,600-hp SNECMA R.210. Small numbers were made of the L-series, with stroke increased to 185 mm, the 14L being a 1,400-hp unit and the 18L a giant 56-litre engine of 1,900 hp, both dropped in 1939. Far more important was the attractive 14M Mars series, with smaller cylinders 122 × 116 mm, capacity 18.98 litres. The Mars was only

Top *The Germans picked the Gnome-Rhône 14M of 700 hp to power the Henschel Hs 129B anti-tank aircraft. This illustration comes from a Luftwaffe training slide.*

Above *Last of the Gnome-Rhône radials, the GR 14U of 1949 was a 44-litre engine rated at 2,200 hp and weighing 1,250 kg. The long gearbox/torquemeter package was configured for a low-drag cowl.*

0.95 m in diameter at first, later versions being about 1 m, and this helped it gain many applications in 1936-40 at ratings from 600 to 710 hp at 3,100 rpm. SNECMA later developed these engines into the 14X with output up to 820 hp at 3,400 rpm, an exceptionally high speed for a large radial.

Green (UNITED KINGDOM)

Gustavus Green, who died in 1964 at 99, was unrelated to Major F.M. Green of the Royal Aircraft Factory. He built his first lightweight 4-in-line water-cooled engine in 1905. His patents were administered by an office in Berners St, London, and manufacture was handled by the Aster Engineering Company. All his engines were seemingly massive water-cooled in-lines with shining copper water jackets, held on the steel cylinders by rubber seals which part-vulcanised with the heat. Two valves per cylinder were worked by an overhead camshaft and rockers, the gear being enclosed from 1910 on. Other features included force-feed lubrication, HT magneto ignition, flywheel and four large bolts to hold down each cylinder and, in most cases, also secure the phosphor-bronze/white metal bearings of the crankshaft. Green won both the British government prizes for an aero engine, £1,000 in 1909 and £5,000 in 1914, with his two main production engines. These were the 30-35, rated at 30 hp at 1,100 rpm and 40 hp at 1,220 rpm, with cylinders 105 × 120 mm and weighing 150 lb, and the 50-60, rated at 50 hp at 1,050 rpm and 70 hp at 1,200 rpm, with cylinders 140 × 146 mm and weighing 259 lb. Green made a 100-hp V-8 and in December 1911 impressed everyone with his 6-in-line, with two Zenith carburettors each feeding three 140 × 146 mm cylinders and rated at 82 hp. Like all Greens it was beautifully made and reliable, but at 450 lb the weight was a problem and wartime production went into fast boats. The 275-hp V-12 of 1915 remained a prototype.

The Green 6-cylinder of 1914, which was rated at 120 hp. Cylinders were 120 × 152 mm, and the whole engine was robust and simple. Crankcase ventilators were styled after those of ocean liners!

Guiberson (USA)

Guiberson Corporation of Dallas made oil-industry equipment, and from 1928 researched diesel aero and automotive engines. In 1932 Guiberson Diesel Engine Company was registered, and Austrian F.A. Thaheld had by that time not only designed the A-980 9-cylinder diesel radial but had begun flight testing in a Waco. Rated at 185 hp at 1,925 rpm, the A-980 took the Waco 960 miles to Detroit on 96 gallons of furnace oil at 7c a gallon. An Approved Type Certificate was awarded in November 1931. Next came the A-918, under chief engineer C.C. Spangenberger, rated at 253 hp and sold to the Navy in 1934. Last was the A-1020, for which an ATC was awarded in February 1940 and which flew over 1,000 h in a Reliant. This had nine 5.125 × 5.5 in cylinders, 1,021 cu in, and for a weight of 653 lb gave 310 hp at 2,150 rpm, with sfc of 0.382. The 1020 started easily with a Coffman cartridge starter and was flown to 18,500 ft in the Reliant (well above normal ceiling) whilst still climbing strongly. Buda Corporation planned to make it, but during World War 2 Guiberson had to make tank engines and subsequently was never able to get enough orders for the otherwise excellent 1020.

Halford (UNITED KINGDOM)

Major Frank Bernard Halford graduated from Nottingham, became a flying instructor at Brooklands, was appointed an engine examiner at the newly created AID (Aeronautical Inspection Directorate), was posted to France in the RFC and in early 1916 was recalled to produce what became the BHP. After the war he did important work with Harry Ricardo in Britain and the USA and then headed the engine section of Airdisco where he redesigned the Puma into the Nimbus, with capacity of 20.7 litres and rpm increased to 1,600, using the stronger crankshaft of the Galloway Atlantic; it was rated at 335 hp. He also redesigned the big Fiat A 12. He designed all the Cirrus engines, followed by all the Gipsy engines. In 1928 he agreed to design also for Napier, producing the Rapier, Dagger and Sabre, finishing with the de Havilland gas turbines.

Hall-Scott (USA)

With Curtiss, this San Francisco company was the US leader in water-cooled engines prior to World War 1. The first (pre-1908) engine was a 4-in-line supposed to give 30 hp, with cylinders 4 in square. Subsequent engines were mainly 6-in-lines or V-8s, one of the latter (designated an A-2) rated at 60 hp at 1,100 rpm flying at the 1910 Belmont Park meet. The A-3 had stroke increased to 5 in, giving 80 hp, whilst the A-4 was the corresponding 6-in-line. The A-5 of 1912 was a V-8 built in quite large numbers, the camshaft being moved from the top of the crankcase to a position above each cylinder block; rating of this enlarged (5 × 7 in cylinders, 1,100 cu in) engine was 125 hp at 1,250 rpm. Smaller numbers were made of the 4-in-line 100-hp A-7 and the corresponding 200-hp V-8. The big 5 × 7 in cylinder was the starting point of the Liberty.

Heinkel (GERMANY)

In February 1936 Professor R.W. Pohl of Göttingen persuaded Ernst Heinkel to hire one of his post-graduate students, Hans-Joachim Pabst von Ohain. A few months earlier von Ohain had obtained a patent for his turbojet, with an axial-plus-centrifugal compressor, annular combustor (patented by fellow-student Max Hahn) and inward radial turbine. In later years von Ohain, who stayed in the USA after being sent there in Operation Paperclip in 1945, denied that he knew anything of Whittle's work, though his colleague Dipl-Ing Wilhelm Gundermann (who worked with him at Heinkel from the start in April 1936) has recorded, 'We kept fully up to date with such patents as Whittle's and Milo AB in Sweden'.

Heinkel had no engine facilities or staff, but let his new protégé construct a simple rig to prove whether the idea worked. It was in no sense a turbojet but a demo rig burning hydrogen gas to avoid the problems of liquid fuel atomization. Called the HeS (S for *strahl* = jet) 1, it was ready in March 1937 and eventually gave about 250 kg thrust. No company wanted to assist with burning liquid fuel at such intensity, and Heinkel was forced to gather a small research staff at Rostock-Marienehe to help design the HeS 2, the first flight engine.

As built in 1938 this comprised an enlarged axial/centrifugal compressor, now with a ball-bearing between the stages, delivering to a flow divider, one flow diluting the flame and the other passing through 16 so-called partial chambers separated by radial walls. Each of these contained a vaporiser tube heated by an externally suplied hydrogen flame; once the 16 vaporisers were glowing the liquid fuel was turned on and the gas disconnected. The burning length had to be many times greater than in the original rig and a

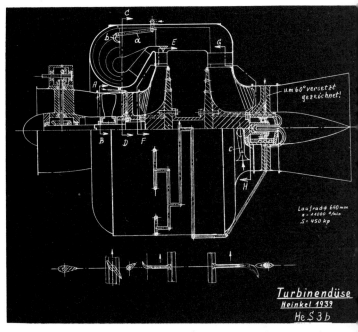

Dipl-Ing Wilhelm Gundermann's original rough drawing of the Heinkel (or Heinkel-Hirth) HeS 3B, the first turbojet to propel an aircraft in August 1939. The only feature seldom seen today is the inward-radial turbine.

large folded combustor was used, the inward-radial turbine having to be moved downstream from the compressor. As originally built it gave less than 20 per cent of the design thrust of 500 kg, and several rebuilds were needed to provide more room for a larger compressor and much larger combustion system, all of which considerably increased the engine's diameter. The result was the HeS 3, the first example of which, designated HeS 3A, was air-tested slung beneath He 118 V2 (*D-OVIE*) in about May 1939. The engine had a diameter of 1.2 m and weighed 360 kg. As first run at the start of 1939 the 3A gave about 400 kg thrust, but this increased by various improvements to 450 kg at the time of first flight, this figure falling to 370 kg at 200 km/h and 345 kg at 400 km/h, in all cases at 11,000 rpm.

Meanwhile the He 178 aircraft had been designed around the engine, which for this application was designated HeS 3B, with a long inlet duct and rear jet-pipe. These additions reduced thrust to a best figure of 380 kg (838 lb). Erich Warsitz began He 178 taxi trials (with a brief hop) on 24 August 1939, followed by a full flight on 27 August. This was the world's first jet flight, though the 178 flew seldom thereafter. It was a very defective aircraft and was limited to 300 km/h even with the later S6 of 450 kg thrust.

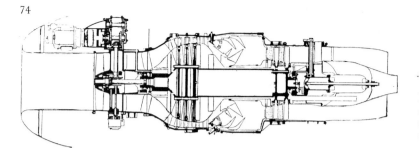

Left *The Heinkel-Hirth HeS 109-011 turbojet had an axial 'inducer' stage at the inlet followed by a diagonal axial/centrifugal stage and then 3 axial stages.*

Below *The HM 504 was the basic inverted 4-in-line of Hirth's family of beautifully made air-cooled engines which culminated in the HM 512A inverted V-12 of 400 hp.*

In early 1939 Heinkel's Robert Lüsser had gathered a team to design the He 280 twin-jet fighter, to be powered by two HeS 8A engines. Heinkel had not informed the RLM (German Air Ministry) of his jet work until late 1938. He received no support until late 1939; even then he was discouraged, as an aircraft manufacturer with no engine facilities. However, believing he was well ahead with something very important, Heinkel was determined to press ahead, even with his own money and in a climate where good engineers were almost unavailable. To avoid the losses caused by long ducts the He 280 had underwing engines, and in turn this forced the HeS 8A to be of small diameter. Redesignated 109-001, it matured at only 775 mm diameter and a weight of 380 kg, but was nowhere near the hoped-for thrust of about 480 kg when the He 280 was flown on 2 April 1941.

By this time Heinkel had hired Max Adolf Müller from Junkers to develop the 109-006 axial turbojet, was trying to build various ducted-fan engines, and was fast falling behind Junkers and even BMW in timing. He had, however, become chief shareholder of Hirth, and took over part of that company's plant in late 1941 to expand his engine operations. This enabled him to scrap earlier projects and with RLM backing undertake a completely new and important turbojet, the Heinkel-Hirth 109-011. Von Ohain was in charge, and development got under way in September 1942; the 001 was dropped, having reached about 590 kg thrust. The 011 had a diagonal compressor followed by 3 axial stages, a 16-burner annular combustor with turbulence 'fingers', 2-stage turbine with hollow blades, and a nozzle with a sliding bullet which was retracted only for idling. Diameter was 875 mm, weight 950 kg, and design thrust 1,300 kg at 10,000 rpm. Ten 011s ran, the first in September 1943, and one flew slung under a Ju 88. The corresponding 109-021 turboprop was never completed. As for the 006 turbojet, this made amazing progress but was dropped in early 1943 because of the (probably mistaken) RLM belief that an engine of 617 mm diameter could not be sufficiently powerful for combat aircraft.

Hiro (JAPAN)

The chief engine made by this aircraft firm was the Type 91, a water-cooled W-12 derived from the Napier Lion, rated at 620-750 hp. It was in production in the first half of the 1930s.

Hirth (GERMANY)

Hellmuth Hirth was a famous pre-1914 aviator and no mean engineer. During World War 1 he planned an improved form of engine with roller bearings throughout, including big and little ends of the conrods. In turn this meant a multi-section crankshaft, and brother Dr Albert Hirth patented the Hirth coupling now used on many gas turbines to ensure automatically perfect assembly of adjacent sections. Other features included cast-iron cylinders with light-alloy heads, pushrods with ball-ended tappets driving the overhead valves via needle-roller rockers, very small metered oil flow to the thrust bearing and cylinders (with oil splash to other bearings), carburettor air taken warm from the crankcase (cast in Elektron magnesium alloy) and an overall exceptional standard of finish.

The first engine to be marketed, the HM 60 inverted 4-in-line of 65 hp, ran in June 1923 and was initially sold in 1924 by the Versuchsbau Hellmuth Hirth at Stuttgart-Zuffenhausen. Despite its almost mistakenly high quality it did so well, in particular winning the Rundflug and eight other major sporting events from 1927 onwards, that the factory expanded and a new company, Hirth-Motoren GmbH, was founded in 1931. The HM 504, 506 and 508 quickly appeared, with inverted 4, 6 and V-8 layout, with cylinders 105 × 115 mm. The 504 was mass-produced at 105 hp at 2,530 rpm, the 506 was rated at 160 hp and the 508 at 280 hp at 3,000 rpm. A few HM 512 inverted V-12s rated at 400 hp were sold after 1938. Hellmuth died in June 1938, and to give Heinkel an engine team and floorspace the RLM tried to arrange for a take-over by the aircraft firm in May 1941. This was blocked by Messerschmitt but eventually was forced through by the end of the year, subsequent Heinkel jet engines being known as Heinkel-Hirth.

Hispano-Suiza (SEE TEXT)

Assigning a nationality to this company is difficult. The name means Spanish-Swiss, because Marc Birkigt (1878-1953) was Swiss but in 1904 set up his great factory at Barcelona, chiefly to make cars of the highest quality. In April 1911 he formed the Soc. Hispano-Suiza at Levallois-Perret, Paris, soon moving a few kilometres to Bois-Colombes where the French factory soon far outgrew its Spanish original.

In 1915 he began production of his first aero engine, and from this stemmed engines which by 1918 had been licensed to Aries, Ballot, Brasier, Chenard & Walcker, Delaunay-Belleville, De Dion Bouton, Doriot-Flandrin-Parant, Fives Lille, Leflavie et Cie, Mayen, Peugeot, SCAP and Voisin in France; Wolseley Motors in Britain; Itala, Nagliata and SCAT in Italy; Wright-Martin in the USA; H-S itself in Spain; the National Arsenal in Czarist Russia; and Mitsubishi in Japan!

Clearly Birkigt's engine was important, and yet, while it was ahead of all others in some respects, it made demands on its builders which many licensees could not meet, and this caused prolonged difficulty. The original November 1914 engine, the Type A, was a water-cooled V-8 with cylinders 120 × 130 mm, capacity 11.76 litres. It weighed 202 kg and was rated at 140 hp at 1,400 rpm. Its most striking feature was that the two cylinder blocks, set at 90°, were monobloc aluminium castings. These gave adequate strength to the thin but deep crankcase which contained all oil not going round the unprecedented force-feed system to all parts of the engine.

Each cylinder block was stove-enamelled to combat porosity, and into each were screwed machined steel cylinder liners threaded over their entire length, with closed tops into which the valves seated. Radial shafts

Geared Hispano-Suiza V-8s in production in England in 1917. Such engines gave serious trouble in the SE.5a programme, but the basic design was superb.

at the rear drove the overhead camshafts, the valve gear being totally enclosed. A vertical shaft at the rear drove the oil and water pumps, and a cross-shaft the two magnetos. The Zenith or Claudel carburettor sat between the blocks feeding via bifurcated pipes and then through passages cast in the blocks. Similar passages carried out the exhaust on the outer sides, and it was to be a trademark of Hispano engines that the intermediate cylinders in each block were to have exhausts grouped into pairs, so that a 6-cylinder block resembled letter 'P' in Morse. Fork-and-blade conrods were used, running in plain bearings on a short crankshaft without balance weights, the latter running in 4 plain bearings and a rear ball bearing.

Overall the Hispano A was the most promising engine in production in 1915. It was short and compact compared with the big 6-in-lines, had lower frontal area than a rotary and set an excellent figure for specific weight. The production A gave 150 hp at 1,400 rpm, and a few Type B 4-cylinder in-lines were made weighing 143 kg and giving 75 hp. By December 1915 the developed 8Aa was in production, rated at 175 hp at 1,700 rpm, soon followed by the 200-205 hp 8Ab cleared to 2,000 rpm and 8Be rated at 215 hp at 2,150 rpm. The 8Ba of 1916 introduced the geared drive, rated at 220 hp at 2,150 rpm. The Wolseley W.4A Python was a copy of the Type A, but heavier at 455 lb. The W.4A Viper had higher compression (5.3 instead of 4.8) and gave 200-210 hp, while the W.4B Adder had a balanced crankshaft and reduction gear. Wright-Martin's Simplex Model E was very like an 8Aa, rated at 180 hp at 1,750 rpm. In general a Hispano-built engine was a high-quality product which, if maintained with great care, would give high and reliable performance for about 20 h between overhauls. Less successful attributes were the poor cylinder cooling, distortion of the thin exhaust valves, and a series of major problems with geared engines including uneven heat-treatment of the helical pinions and fatigue failures of the crankshaft. From summer 1917 there was a crisis in SE.5a deliveries because Wolseley crankshafts seldom lasted more than 4 h. As a stop-gap the firm was told to switch to the direct-drive engine; instead of rushing through the 1,100 engines, Wolseley went back to the drawing board to get the same 200-plus horsepower from the high-compression Viper. Eventually good geared engines were produced, and some were fitted with Hispano's patented *moteur canon* in which a gun (up to the massive 37-mm Puteaux) was mounted between the cylinder blocks to fire through the hub of the propeller. Guynemer's SPAD XII, in which he gained four victories with the 37-mm gun, first flew on 5 July 1917.

In December 1916 the Model H introduced cylinders 140 × 150 mm, capacity 18.47 litres, rated at 308 hp at 1,850 rpm for a weight of 270 kg. By this time the original geared engines were giving up to 240 hp, and the bigger model was developed to 340 hp. This was later built as the Wright-Hispano. Total output of all models in 1915-18 is believed to have been 49,893, comprising 12,593 of 150-180 hp, 28,977 of 200-220 hp and 8,323 of 300 hp.

After the war the main aero-engine factory continued to be Bois-Colombes, in a road renamed for Captain Guynemer. Here Hispano-Suiza gradually improved their range so that they required less constant attention and became amongst the most reliable of all engines, setting numerous speed and distance records. By 1925 the most important products were 60° V-12s, but there were 6-in-lines, W-12s and even W-18s, the commonest cylinder size being 150 × 170 mm, giving capacity of 36.05 litres for a 12-cylinder engine. By far the most important family was the 12Y of 1933, some 18,000 of which were made at ratings from 800 to over 900 hp, the commonest fighter engine being the 860-hp 12Y-31. Compared with the Merlin of the same period the 12Y engines were of larger capacity (similar to the Griffon), lighter and of lower power, largely because they were limited to 2,400 instead of 3,000 rpm.

The most powerful mass-produced engine was the

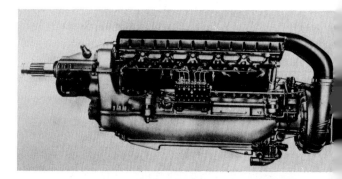

Below *One of the last of the pre-war Hispano-Suiza V-12s, the HS 12Z was giving 1,800 hp in early 1940, with Turboméca supercharger and direct injection.*

Bottom *Spurred by the threat of Arsenal with its monsters derived from the Jumo 213, Hispano-Suiza went mad as well and produced the 24Z, with four 6-cylinder blocks of 12Z type.*

12Y-45 with Szydlowski (Turboméca) supercharger, rated at 910 hp. The 1,100-hp 12Y-51 was entering production at the time of the capitulation, and the 1,280-hp 89ter powered the Arsenal VG 39 fighter of February 1940. The 12Z with direct fuel injection ran at 1,800 hp at 2,800 rpm before the capitulation, and powered the D.520Z in February 1943.

During the war the design of an H-24 engine with four 12Z cylinder blocks was slowly completed, and this big engine was tested in 1949-52. It had four Lavalette injection pumps, dual 2-speed superchargers on a transverse shaft, and each half engine drove half the contra-rotating propeller. The 24Z gave almost 4,000 hp.

The war interrupted design of the 12B, but this ran in 1945 and emerged in several forms in 1947 with ratings up to 2,200 hp. Still using the 150 × 170 mm cylinders, the 12B was mechanically a completely different and much stronger engine, but it was too late. In 1938 a French equivalent of the German HZ-anlage had been the use of a fuselage-mounted HS 12Xirs of 690 hp driving a 3-stage compressor to supercharge the two HS 12Y-32/33 wing engines of the NC 150 (flown for the first time on 11 May 1939) which were rated at 955 hp at 9 km (29,530 ft).

In 1929 the company bought a licence for the Wright Whirlwind, and subsequently produced about 2,500 radials with various numbers of cylinders, the most important being the 14AB, an extremely compact 14-cylinder unit rated at 650 hp in 1934 and up to 800 hp in the 14AB-12/13 of late 1939. This family are often confused with the bigger 14AA of 1,100-1,350 hp. Hispano also built small numbers of HS 9V and related 9-cylinder engines of greater diameter than the 14s, rated around 600 hp, but none of the company's radials (all using Wright features) made much impression on the GR-dominated market. In 1934 the HS 12Y was licensed to the Soviet Union, and, as the M-100, became the baseline engine for the VK series developed by V.Ya. Klimov (*qv*).

In 1946 Hispano-Suiza obtained a licence for the Rolls-Royce Nene turbojet, and large numbers were made of Mks 101, 102, 104 and 105, as well as prototypes of the afterburning Mk 102B. Hispano itself developed the R.300 with increased airflow and hollow air-cooled turbine stators, rated at 2,700 kg. This was dropped in 1951 in favour of a licence for the Tay 250 of 2,850 kg thrust; the afterburning 250R of 3,850 kg thrust remained a prototype. The final derived engine was the Verdon 350, first flown in a Mystère in August 1953, in which further increases in mass flow, rpm and temperature increased thrust to 3,500 kg at 11,100 rpm, weight being 935 kg. The company participated in the Avon installations in the Caravelle airliner and Dassault fighters, and in 1954

designed a small axial turbojet, the R.800, for light fighters. Rated at 1,800 kg at 12,200 rpm, this had a 7-stage compressor, diameter of 692 mm and weight of 303 kg. It was dropped in early 1956. Hispano-Suiza then diversified, later engine work including production for SEF of the SEPR 844 rocket for the Mirage III, and licence-assembly and participation in manufacture of the RR Tyne. In December 1968 the company was acquired by SNECMA.

Hitachi (JAPAN)

This large conglomerate made aero engines from about 1929, and concentrated on low-powered engines up to the collapse in 1945: the GK2 Amakaze 9-cylinder radial of 340-610 hp, the Ha-12 and Kamikaze 7-cylinder radials of 150-160 hp, the Ha-13, Ha-42 and Tempu 9-cylinder radials of 310-510 hp, and the GK4 and Ha-47 inverted 4-in-lines of 100-110 hp.

I

IAE (MULTINATIONAL)

International Aero Engines AG is a company incorporated in Switzerland with operating HQ at 287 Main St, East Hartford, Connecticut, USA, near Pratt & Whitney. Its members are that company and Rolls-Royce (each 30%), JAEC (23%), MTU (11%) and Fiat (6%). JAEC, Japan Aero Engine Corporation, is formed by IHI, Kawasaki and Mitsubishi. Seldom have such multinational resources been applied to a single product, this being the V2500 commercial turbofan in the 25,000-lb class, for certification in April 1988. RR had previously collaborated on an engine known as the RJ.500, and this is one of several engines which contributed to the V2500. The latter clearly meets the CFM56-5 in head-on competition. Basic responsibilities are: Japan, fan and LP spool; UK, HP spool; USA, combustor and HP turbine; West Germany, LP turbine; and Italy, accessory gearbox.

Isotov (SOVIET UNION)

General Constructor Sergei Pietrovich Isotov (1918-83) was responsible for two important helicopter turboshaft engines. For the Mi-8 his bureau developed the TV2-117A, and the associated VR-8A

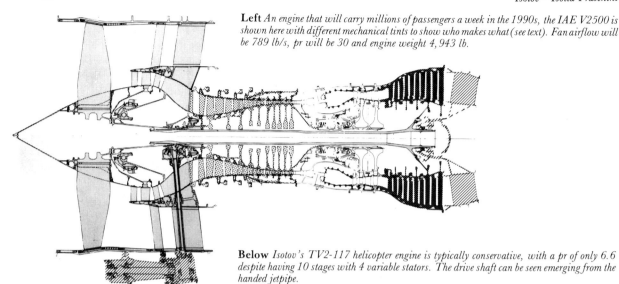

Left *An engine that will carry millions of passengers a week in the 1990s, the IAE V2500 is shown here with different mechanical tints to show who makes what (see text). Fan airflow will be 789 lb/s, pr will be 30 and engine weight 4,943 lb.*

Below *Isotov's TV2-117 helicopter engine is typically conservative, with a pr of only 6.6 despite having 10 stages with 4 variable stators. The drive shaft can be seen emerging from the handed jetpipe.*

gearbox which combines the drive from left and right engines. Very conservative by Western standards, the TV2 has a 10-stage compressor (pressure ratio 6.6 at 21,200 rpm), annular combustor with eight burners, and two-stage gas-generator and power turbines. A large jetpipe turns the gas through 60° to a side exit upstream of the rear output shaft. The engine is 2,835 mm long, weighs 330 kg without the generator or other accessories, and has a maximum rating of 1,700 shp, with sfc of 0.606 lb/h/shp. This engine was in production at 1,400 shp in 1962 and fully rated by 1964. From it stemmed the uprated TV3 series used in Kamov helicopters and later Mil machines. Fitted with a pneumatic starter, the TV3-117MT is rated at 1,900 shp (2½-minute contingency 2,200) while the electrically started civil TV3-117V is rated at 2,225 shp.

For the small Mi-2 Isotov developed the GTD-350 which exactly followed the Allison 250 in layout with 7 axial and one centrifugal compressor stage (pr 6.05 and airflow 2.19 kg/s at 45,000 rpm), twin lateral pipes leading to the single reverse-flow combustor, single-stage gas-generator turbine and two-stage

power turbine, the latter running at a constant 24,000 rpm. Weighing 135 kg, the GTD-350 has a take-off rating of 400 shp with sfc of 0.805 lb/h/shp. Like the helicopter, this engine was passed to Poland for production. Isotov also produced the TVD-850 turboprop of 810 shp which was an unsuccessful candidate engine for the An-28.

Isotta-Fraschini (ITALY)

Like many famous European aero-engine firms, this company began as a builder of high-quality cars, Fabbrica Automobili Isotta-Fraschini being established in 1898. Early aero engines, the first in July 1910, were massive water-cooled 4-in-lines, about 50 of which powered Italian airships, land aeroplanes and seaplanes before August 1914. During the war the company claimed to have delivered 'nearly 5,000 engines' and 'nearly all the aero engines produced in Italy were made under Isotta-Fraschini licence'—a claim which appears to overlook Fiat's total of just over 15,000!

Early engines were prefixed V (*volo* = flight), most being 6-in-lines with cylinders of cast-iron made in

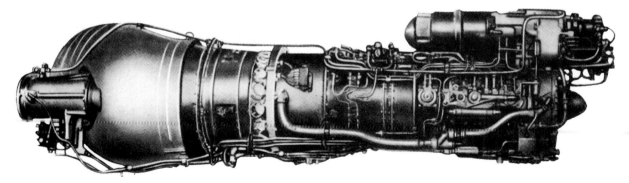

pairs with common heads; bore and stroke varied tremendously, but most models had take-off rpm of 1,800, giving powers from 120 to 280 hp. The V.5 of 1915 was a long straight-8 of 19 litres rated at 200 hp but weighing a massive 351 kg. Easily the most important of all the company's engines was the V.6 of 1917 with 6 cylinders, 140 × 180 mm and giving 250 hp at 1,650 rpm. Many derived engines followed, most having a monobloc cast aluminium head for each bank of cylinders, whilst retaining separate iron cylinders, usually with sheet water jackets added. In 1920 the company collected various improved features and used them in the first Asso (ace), a name used for engines of every conceivable configuration and power including (for example) the Asso 80 (80 hp) 4-in-line and Asso 1,000 (1,000 hp), both of 1928, the latter having 57.26 litres in three banks of 6 cylinders: three push/pull pairs of these monster engines powered the vast Ca 90.

By World War 2 Isotta-Fraschini was firmly in the trap of having a profusion of contrasting engines all made in very small numbers. The only real production was of the Delta RC 35, an inverted V-12 (with air-cooled cylinders introduced in 1927 with the 480-hp Asso Caccia upright V-12 for fighters) of which over 3,300 were made at powers around 750 hp. Other World War 2 engines included the 500-hp Gamma RC 15 inverted V-12, 900-hp Asso L 121RC40 liquid-cooled V-12, 450-hp Astro 7C40 and 890-hp Astro 14C40 7 and 14-cylinder radials, the 1,500-hp Asso L 180RC115 liquid-cooled inverted W-18, and the 1,200-hp Zeta RC 25/60 air-cooled X-24 (almost two Deltas) which ran in December 1941 and powered the Caproni-Vizzola F.6Z fighter.

Ivchyenko (SOVIET UNION)

General Constructor Aleksandr G. Ivchyenko began his career with small piston engines, being permitted to use his own AI designations from 1944. Fair numbers were made of the AI-4 air-cooled 4-stroke flat-4 qualified in 1946 at 52 hp and produced mainly for helicopters at 55 hp. Small numbers were produced of the AI-10 5-cylinder radial, qualified at 80 hp in 1946. A year later the first production version of the AI-14 was qualified at 240 hp. This radial has 9 cylinders 105 × 130 mm, capacity 10.16 litres, and many thousands have been produced at 260 hp, 285 hp (the 14VF for helicopters) and 300 hp (the 14RF).

Since 1952 these have been licensed to Poland, where the chief model built at PZL-Kalisz is the 260-hp AI-14RA, and also developed by Vedeneyev. The more powerful AI-26 has 7 cylinders 155.5 × 155 mm, 20.6 litres, and was qualified in 1946 at 500 hp, subsequently being mass-produced in fixed-wing and vertical installations at up to 575 hp. Licensed to

Above *The Isotta-Fraschini V.6 of 1917 was one of the first engines to have totally enclosed valve gear. It was made in large numbers and also licensed to Bianchi, San Giorgio and Romeo.*

Below *Made in Poland since 1952, Ivchyenko's AI-14RA is started by compressed air piped to the cylinder heads.*

Poland, it has been developed into the PZL-3 (*qv*).

The AI-20 single-shaft turboprop was begun at the Kuibyshyev engine plant under Kuznetsov as the NK-4 with a largely German team; in 1952 tests showed lower sfc than the rival VK-2 and it was adopted for production, being transferred to the even bigger Zaporozhye plant under Ivchyenko as the AI-20. A constant-speed engine (12,300 rpm), it has a 10-stage compressor handling 20.7 kg (45.6 lb)/s and with pressure ratio at altitude cruise conditions of 9.2; the annular combustor has 10 typically Russian conical burner mixers, and the turbine has 3 stages and is overhung. Qualified in 1955 at 3,750 ehp, it was produced as the AI-20K at 3,945 ehp (also made at Shanghai as the Wo-Jiang 6), the 20M at 4,190 ehp, the marinized 20D at 4,190 ehp and the 20DM at 5,180 ehp.

The AI-24 is almost an AI-20 scaled-down to an airflow of 14.4 kg (31.7 lb)/s at constant 15,100 rpm, pressure ratio at max-cruise altitude being 7.85; the combustor has 8 burners of Simplex type, but otherwise the layout is similar to the bigger engine. One of the material differences is that the compressor casing is steel and made in left/right halves instead of magnesium in upper/lower halves. The AI-24 was qualified at 2,515 ehp in 1960, and was followed by the 24A rated at the same power maintained to hot/high conditions by water injection; this is made at Shanghai as the WJ-5A. The 24T is uprated to 2,820 ehp, but was replaced in production in 1980 by the slightly improved 24VT. These engines have been used in widespread propfan testing since the 1970s.

For the Yak-40 Ivchyenko produced the AI-25 turbofan, a robust and conservative engine which achieves a mere 8:1 pressure ratio from a 3-stage fan (1.695) and 8-stage compressor (4.68); there are 12 burners in the annular chamber, a single HP and two-stage LP turbine, and pneumatic starter. Bypass ratio is 2. The AI-25 powered the Yak-40 on its first flight in 1966 but was not qualified until later, take-off rating at 16,640 rpm being 1,500 kg (3,307 lb) with sfc 0.56. The long-jetpipe AI-25TL powers the L 39 trainer, while a variant with air-cooled turbine is

Top left The AI-20M is one of the mass-produced Ivchyenko turboprops, its weight being 1,040 kg. Like several other turboprops it runs at constant rpm.

Above left The AI-24A is bigger and heavier (600 kg) than equivalent Western engines, but has been made in the Soviet Union and China in large numbers.

Left Ivchyenko again went for robust simplicity rather than performance in the AI-25 turbofan, though he tried to reduce weight by using aluminium, magnesium and titanium alloys where possible. This version weighs 290 kg without accessories.

rated at 3,850 lb. On Ivchyenko's death in June 1968 the bureau became led by Lotarev (*qv*).

J

Jacobs (USA)
The Jacobs Aircraft Engine Company was formed in 1930 in Camden, New Jersey, USA, later moving to Pottstown, Pennsylvania. Starting with small 55-hp 3-cylinder engines, it quickly progressed to a 7-cylinder of 150 hp and then in 1933 enlarged this from 589 to 757 cu in to create the L-4, which had an active life of 50 years. Better known today by its wartime designation of R-755, it has steel cylinders 5.25 × 5 in with aluminium-alloy heads incorporating aluminium-bronze valve seats. The earliest L-4 versions were rated at 200 hp at 2,000 rpm, but post-war geared models developed up to 350 hp at 2,500 rpm. One of these R-755E variants was for helicopters. Much smaller numbers were made of the larger 7-cylinder L-5 of 285 hp of 1935 and the L-6 (R-915) of 5.5 in bore and stroke used at 330 hp in various wartime aeroplanes and autogyros. In the 1970s Page Industries of Oklahoma bought all rights and for a few years manufactured small numbers of R-755s and spares, including the turbocharged R-755S.

Jalbert (FRANCE)
Sponsored by the French air ministry, Ateliers et Chantiers de la Loire built prototypes of three types of diesel to Jalbert designs. First to run, in 1928, was a 4-cylinder water-cooled in-line, of 160 hp, followed by a 235 hp inverted 6-in-line and a challenging water-cooled H-16 intended to give 600 hp at 2,400 rpm.

Jendrassik (HUNGARY)
Another of the forgotten pioneers, György Jendrassik built the world's first turboprop, and it was in most respects a sound engine of commendably modern design. A senior engineer at Ganz wagon works, Budapest, he began design of a 100-hp unit in 1932 and ran it in 1937. By this time he had completed design of the Cs-1 turboprop, to be rated at 1,000 hp at 13,500 rpm. Features included a cast inlet housing the 0.119 reduction gear, a 15-stage compressor and 11-stage turbine hung between two bearings and joined by a large-diameter tubular shaft, a folded

Far from being a crude 'breadboard' research rig, the world's first turboprop was a compact well-designed engine. Here is the Jendrassik Cs-1 ready for test in Budapest in August 1940.

reverse-flow annular combustor, air-cooled turbine discs and extended-root blades. The Cs-1 ran in August 1940 and was so promising that design of the X/H (also called RMI 1) twin-turboprop fighter/bomber went ahead (under Varga at the RMI). The whole project was killed by the war situation in 1941, and supply of DB 605 engines of higher power.

Junkers (GERMANY)

Professor Hugo Junkers set up a factory making small marine diesels in 1913 (two years before he made aircraft), and in the same year built a rather primitive aero oil engine (using spark ignition), the MO-3 4-in-line, followed by the MO-8 6-in-line of 1914. In 1916 he ran the giant FO-2 rated at 500 hp at the high speed of 2,400 rpm, and even designed the corresponding V-12 to be rated at 1,000 hp, though this was not built.

After World War 1 Junkers decided to develop ordinary petrol engines for his own aircraft, and for possible sale to others, and quickly produced a series of conservatively rated water-cooled in-line engines, notable chiefly for having ball-type main bearings and two exceptionally large valves per cylinder. The L1 of 1921 was a 6-in-line rated at 80 hp, and the bigger 6-cylinder L2 began life at 195 hp and grew to 220 hp, or in the L2a to 230. The most important of the early engines was the big L5 of 1922, with cylinders 160 × 190 mm, giving capacity of 22.92 litres. Weighing about 325 kg, these massive engines began at only 280 hp at 1,450 rpm, but most production L5s were rated at 310 hp and the L5g at 340. Using the same size of cylinder, Junkers then redesigned the L5 to run at 2,000 rpm and give 400 hp, the result being the L8 of 1923, in which year a separate engine company, Junkers Motorenbau (Jumo) was formed. A very few L55s were built, these being V-12s based on the L5 rated at 600 hp. More important was the

L88 of 1925, the V-12 derived from the L8. The production L88 was normally rated at 800 hp, and the high-altitude L88a (flown in the Ju 49 to over 12.5 km) had advanced superchargers maintaining power to 700 hp at 5.8 km (over 19,000 ft).

By 1924 the oil-engine team had switched to full compression-ignition diesels, using a totally different configuration in which two opposed pistons work in each very long cylinder, crankshafts above and below the engine driving the propeller via trains of gears up the front of the engine. Junkers adopted the two-stroke cycle, one piston uncovering the exhaust ports round the cylinder and the other piston the inlet ports, these being shaped to swirl the air to help expel the exhaust and assist mixing the oil spray injected into the extremely hot compressed air near top dead centre. Such engines tend to be tall and thin, and structural considerations demand massive construction. The first of this important series was the FO3 of late 1925, which had five (in effect 10) giant cylinders and was rated at 830 hp at 1,200 rpm. By 1927 the FO4 had run, using six (12) smaller cylinders and rated at 720 hp at 1,700 rpm. By early 1929 this was mature enough to be flown in the aircraft company's F 24 testbed. This massive engine was put into production as the Jumo 4, later renumbered 204, rated 770 hp at 1,800 rpm and used by Lufthansa in the Ju 52 and re-engined G 38.

In 1932 the cylinder size was reduced yet again, to 105 × 160 mm (160 being the stroke of each piston), giving capacity of 16.62 litres. The resulting Jumo 205 was cleared to run at 2,200 rpm, giving 600 hp, and the wartime 205D with pressure-glycol cooling was rated at 700 hp at 2,600 rpm. The exhaust at barely 500°C made it easy to install a turbocharger, and the wartime Jumo 207 added both a turbo and a gear-driven supercharger, giving 1,000 hp at 3,000 rpm for take-off and 750 hp at 2,800 rpm at up to 12.5 km. The final wartime engines were staggering in concept, the Jumo 223 being a family of related 'box' engines comprising four opposed-piston diesels in one unit with a crankshaft at each corner; weighing 2,370 kg, it gave 2,500 hp at up to 6,000 m. In 1942 it was dropped in favour of the Jumo 224, which was even bigger and designed for 4,500 hp.

In 1933 the Jumo firm wisely embarked on completely new high-power petrol engines, starting with the Jumo 210 and 211, both of which ran in 1936. These were inverted V-12s cooled by Glysantin (glycol), but with different sizes of cylinder, the 210 being 124 × 136 mm (19.7 litres) and the 211 cylinder being 150 × 165 mm, giving 35 litres. Both were fitted with carburettors, and had similar construction with the entire crankcase and cylinder blocks being a single high-quality light-alloy casting. The steel cylinder liners were tightened into the one-piece head

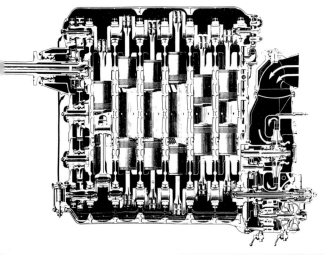

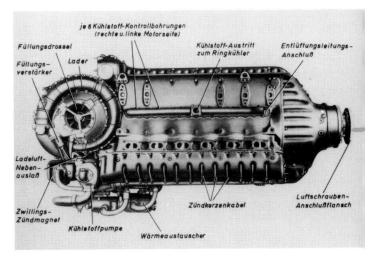

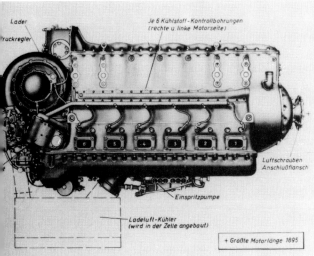

Above left *A Junkers drawing showing the arrangement of the opposed pistons and crankshafts in the Jumo 205 two-stroke diesel. White arrows show airflow into the blower used for scavenging.*

Above *In many respects an outstanding engine, the Jumo 213A-1 powered the Fw 190D-9 at a rating of 2,240 hp with MW50 injection. Similar engines were produced in France after the war.*

Left *The Jumo 211 J was produced in large numbers for the Ju 88. Supercharger was on the right, unlike the DB engines. Photo from Luftwaffe handbook.*

Below left *The Jumo 207B-3 was the ultimate production model of two-stroke diesel. Its climbing power of 750 hp was maintained from sea level to almost 40,000 ft.*

by 4 wet bolts, 14 studs then being used to tighten each 6-cylinder assembly on the main block. The 210 began at 600 hp and reached 730 hp at 2,600 rpm in the 210Ga fitted to the Bf 109C. The bigger 211 was first tested in a Ju 87A, and 68,000 were built in World War 2 at ratings from 1,000 to 1,530 hp (one giving 1,380 hp at 10 km), almost all production versions having direct fuel injection. The 211 had one exhaust and two inlet valves per cylinder, and the supercharger was on the right side at the rear. A typical weight (211A) was 640 kg.

At the start of World War 2 Junkers was engaged in refining the 211 into the 213, a particularly advanced engine with all mechanical features redesigned for continuous running at 3,000 rpm and take-off at 3,250. Typical of the refinements was the addition of a steel flywheel at each end of the underhead camshafts. The basic 213A was rated at 1,776 hp, for a weight of 920 kg, while the 213J, with 4 valves per cylinder, was rated at 2,600 hp at 3,700 rpm, a remarkable speed for so big an engine. Deliveries of 213s reached 9,000. In 1939 work was also well

advanced on the totally new Jumo 222, planned for large-scale production but destined to fly in prototypes only. It was a liquid-cooled radial, with 6 banks each of four 135 × 135 mm cylinders (46.5 litres). It had many unusual features, and the 222A series weighed about 1,080 kg and gave 2,500 hp at 3,200 rpm. The A and B Srs 2 had bore increased to 140 mm, and the C and D had cylinders 145 × 140 mm and gave 3,000 hp at 3,200 rpm.

This MoD(PE) photo shows a Jumo 222A tested at Farnborough in April 1945. At far left is one of the rectangular air inlets, ahead of which is one of three supercharger delivery pipes, which branches to feed Nos 6 and 1 blocks. Above the supercharger is one of the two 24-cylinder magnetos, driven off No 2 block camshaft. Weight was 2,394 lb.

Gas turbines

Back in July 1936 the engine and aircraft companies had been amalgamated, though the two divisions carried on as before. The main board under Koppenberg supported the brilliant and highly political head of airframes, Herbert Wagner, who thought the engine team under Otto Mader too conservative and slow. Before the amalgamation, in April 1936, Wagner started a secret cell at Magdeburg developing a gas turbine, and by 1938 this effort (unknown to Mader) included a profusion of turbojets, turboprops and even stationary plant. The team leader, Max Adolf Müller, managed to get the Jumo 109-006 axial turbojet on test by autumn 1938. This slim (617 mm) axial engine, with pressure ratio 2.9 from 5 stages, had much potential though it was only just able to run under its own power (the Magdeburg group began to switch their effort to diesel-driven ducted fans!). Meanwhile, the engine division at Dessau under Mader continued studies, started as early as 1933, of

turboprops and free-piston gas turbines, and in August 1938 put Anselm Franz in charge of a survey of all gas turbine prospects, accepting an RLM (Air Ministry) study contract a month later. After detailed weight estimates of a free-piston jet engine Franz accepted that only a simple turbojet would be sufficiently light. In April 1939, after acrimonious scenes, the RLM welcomed the engine division's take-over of the Magdeburg work, Wagner having already arranged another job and Müller and half his staff going to Heinkel. By the summer Junkers had an RLM contract for a bigger and conservatively rated turbojet, the 109-004 (the 006 went to Heinkel). Mader insisted on the main effort staying on piston engines, but by 1942 Franz had managed to gather a staff of some 500.

The 004 was designed for a thrust of 700 kg at 900 km/h, and despite much greater familiarity with centrifugal compressors Franz chose an 8-stage axial of 3.1 pr, 6 combustors, single-stage turbine, and jet-pipe arranged for afterburning and with a nozzle bullet to maintain constant gas temperature. Encke designed the compressor blading (as he did for the

BMW 003) and the AEG company assisted with the turbine, which at first had solid blades. The 004A ran in November 1940, but many snags took until early 1942 to overcome. No 5 engine reached 1,000 kg thrust, its weight being 848 kg; on 15 March 1942 another 004A flew slung under a Bf 110, and on 18 July 1942 two A-0 engines powered the Me 262 V3 on an extremely successful flight. About 30 A-series

Below *The Jumo 004B series were started by this Riedel flat-twin 2-stroke (10 hp at 10,000 rpm) which itself has powered karts and microlights.*

Bottom *Sir Roy Fedden took this picture of Jumo 004 Bs wrecked by bombing at Magdeburg, and wrote on the back 'The toll of the orthodox' because the bombers had piston engines.*

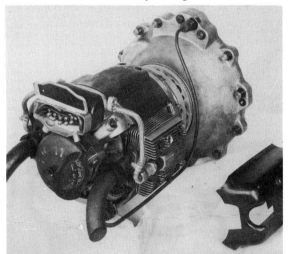

engines were built, followed in 1942-43 by a few B-series in which strategic materials were cut by half, partly by using sheet instead of castings, reducing weight to 748 kg and greatly reducing man-hours and price. The B-1 introduced a compressor with improved blades in the first two stages, raising thrust to 900 kg. This ran in May 1943 and powered the Me 262 V6 in early November. Production on a massive scale was then arranged, the first B-1s becoming available in March 1944. It had a light-alloy compressor of spigoted disc construction with wrapped-sheet stators, aluminised mild-steel combustors burning diesel oil, 61 solid Ni-Cr steel turbine blades soldered and pegged into the martensitic steel disc, a jetpipe bullet positioned by a servo-motor and rack/pinion, and a 10-hp Riedel 2-cylinder 2-stroke starter in the nose bullet (which after the war was popular for go-karts). After much research hollow turbine blades were perfected, and these entered production in December 1944 in the B-4. The engine was still conservatively rated, and in early 1945 production was about to switch to the D-4 rated at 1,050 kg. Roughly 5,000 Jumo 004B engines were built (today's MBB claims over 6,000), TBO being 30 h.

Junkers had a contract for a bigger and more advanced turbojet, the 109-012, and the corresponding 022 turboprop. Planned for the Ju 287, the 012 had an 11-stage compresssor and 2-stage turbine, was to weigh 2,000 kg and give 2,900 kg thrust. The first 012 was never completed.

K

Kawasaki (JAPAN)

This famous company was the fourth largest in Japan in terms of engine production in World War 2. From 1931 it made over 2,000 Ha-9 water-cooled V-12s rated at 710 to 950 hp. Its effort was concentrated on liquid-cooled engines, Kawasaki having produced many thousands of water-cooled BMW VI and derived engines in 1927-40. In 1939 a licence was obtained for the DB 601, and this was redesigned at the Akashi plant to meet local requirements, going into production as the Ha-40 in November 1941.

By 1942 work was well ahead with the 1,500-hp Ha-140, but this was persistently down on power and unreliable. The Ha-201 was an installation of two Ha-40s in the Ki-64 fighter, one ahead of the cockpit driving the rear (fixed-pitch) unit of a contraprop and the other behind the cockpit with a long shaft to the constant-speed front propeller. This extraordinary machine flew in December 1943. Since 1953 Kawasaki has overhauled and licence-built many gas turbines, and in 1981 the company ran the first KJ12 small turbojet, rated at 150 kg thrust for a weight of 40 kg and intended for RPVs and small sporting aircraft.

Kinner (USA)

Kinner Airplane & Motor Corporation was established in Glendale (Los Angeles) in 1919, and began producing a 5-cylinder radial. From 1931 it also made light aircraft, but it went bankrupt in 1937. In its place came Kinner Motors of 1939, and this enjoyed major wartime production, its engines clattering roughly but reliably in thousands of trainers. The K-5 series had cylinders 4.25 × 5.25 in, 372 cu in, and at 100 hp were a close equivalent to Shvetsov's M-11. The B-5s had bore 4.625, 441 cu in and 125 hp, and the R-5 and derivatives (such as the R-55) went to 5 in, 540 cu in, and 160-175 hp. All looked spiky because the cylinders (with widely splayed valve gear

Many hundreds of wartime Kinners are still being lovingly maintained; this is a 160-hp R-55.

driven by 5 camshafts) were bolted to raised platforms on a tiny crankcase drum. In 1944 Kinner launched a very modern flat-6 of 225 hp (geared, 250 hp) which in contrast ran like a sewing machine, but the firm went swiftly downhill from 1945.

Klimov (SOVIET UNION)

Vladimir Yakovlyevich Klimov was one of the first specialists on water-cooled aero engines in the Soviet Union, and played a major role in developing the M-17 from the BMW VI, working wth Mikulin. In late 1933 the GUAP (aero industry directorate) picked the Hispano-Suiza 12Y as a major engine for fighters, and Klimov was instructed to open a KB (construction bureau) to develop it beyond the original licensed engine. The latter was designated M-100, but in December 1940 under the revised designation system for all aviation items Klimov was permitted to use his own initials, which are used here exclusively.

The baseline engine had 2 mm shorter stroke than the French engine, giving capacity of 35.09 litres, and despite many confusing reports to the contrary this never altered in subsequent engines. Production engines of 1935 had single-speed superchargers and a rating of 750 hp, the 100A of 1936 reaching 860 hp at 2,400 rpm. The VK-103 with 2-speed supercharger was qualified in January 1937 at 860 hp, the 103A reaching 960 hp in 1939 and later 1,100 hp on 100-PN fuel. The most numerous model was the VK-105 cleared to 2,700 rpm, made in vast numbers with *moteur canon* and various refinements, the 105 being rated at 1,050 hp and sub-types being the 105P of 1,100 hp (sometimes with TK turbosupercharger maintaining power to 4 km), the PF of 1,260 hp, PF-2 of 1,280 hp and RA of 1,100 hp. VK-105s accounted for 101,000 of the total of over 129,000 of all these V-12 engines in 1935-47. The 1,200-1,350 hp VK-106 remained troublesome prototypes. The VK-107 introduced an air scavenge valve to supplement the previous one exhaust and two inlet valves per cylinder, and being restressed to 2,800 rpm was rated at 1,400 hp on 94/95 fuel in 1942, and 1,650 hp on 100-grade fuel in 1943. Unlike earlier Hispano-derived engines the exhaust outlets were evenly spaced. No production was undertaken of the 1,800-hp VK-108 or 2,073-hp VK-109 of 1945.

By 1945 Klimov was a Major-General Constructor, and member of the Academy of Sciences, heading KBs at GAZ (factory) No 45 in Moscow and GAZ-117 in Leningrad, where S.P. Isotov was his deputy. In early 1946 his GAZ-117 team began trying to copy the RR Nene, but in September 1946 arrival of the actual engine in Moscow resulted in instant scrapping of this primitive effort. Klimov went straight to GAZ-45, with a top team of GAZ-117

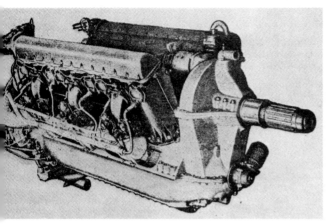

In 1935-46 over 129,000 of Klimov's Hispano-derived engines were produced for Soviet combat aircraft. Most numerous of all was the 1,260-hp VK-105PF.

engineers, copied the British engine in every detail and, designating it RD-45 after the factory, put it into mass production. Klimov directed round-the-clock work on improving the RD-45 and in December 1948 the VK-1 was cleared for production, albeit with a TBO of only 25 h. Airflow was increased from 41 to 45 kg/s, ratings being 2,700 kg, the same as for the VK-1A with different accessory gearbox and longer life, and 3,380 kg for the VK-1F with afterburner.

In late 1949 production began at GAZ-45 and at GAZ-16 at Ufa and GAZ-19 in Kuibyshyev, over 39,000 being produced. Klimov's first gas turbine, the VK-2 single-shaft turboprop, was developed at GAZ-117 from 1947, with 8-stage axial compressor, 7 can-type combustors, 2-stage turbine and planetary reduction gear. In 1950 the VK-2 achieved its design sfc of 270 g/hp/h; this was lower than Kuznetsov's rival NK-4, and Klimov ordered all effort to be switched to reliability. This proved a mistake, and in 1952 official tests showed the NK-4 to have lower sfc and it was picked over the VK-2. The VK-3 was again a totally new design, a single-shaft bypass turbojet directed by S. V. Lyunevich at GAZ-117 in 1952. New features included a supersonic first stage to the axial compressor and an annular combustion chamber. It led to the VK-7. The VK-5, begun at GAZ-117 in 1949 under A. S. Mevius, was the ultimate turbojet derived from the Nene; 10 prototypes were bench and flight tested, with pressure ratio 5.05 and dry thrust of 3,100 kg, but it could not compete with new axial engines.

Koliesov (SOVIET UNION)
A little-known construction bureau (KB), this is reported unofficially to have taken over the assets of

the Dobrynin KB in the mid-1950s. Concentrating initially on large supersonic turbojets, its only identified product is the VD-7, the designation strongly suggesting that it was originally developed under Dobrynin. The VD-7 has always been described as a high-compression single-shaft engine. The first known application was the Myasishchyev M-50 and M-52 series supersonic bombers, and it has long been assumed that a later version powers the Tu-22 (Blinder). Afterburning thrust is at least 14 tonnes (30,865 lb). In 1973 it was officially stated that the Koliesov KB was developing a variable-cycle engine for the Tu-144D SST, functioning as a high-compression turbofan for take-off and subsonic flight and as a turboramjet in the high-supersonic cruise regime; it was said to meet noise/emissions legislation, to be '50 per cent more economical than the NK-144' and to be ready for production, but nothing has been heard since. Koliesov has been named as working in the field of VTOL lift jets, and is believed to have produced the ZM lift engines of the Yak-38 (Forger).

Kossov (SOVIET UNION)
M.A. Kossov was an assistant to Shvetsov in the early 1930s, and in about 1935 was permitted to carry out his own developments of the M-11. Known products include the MG-11 of 1937 rated at 165 hp, the MG-11F with ratings up to 180 hp, the 7-cylinder MG-21 of 1938 rated at 200 hp, the 9-cylinder MG-31 of 1938 rated at 300 hp, and the MG-31F of 330 hp. Designation MG-40 was apparently a refined M-11 of 1934, rated at 140 hp.

Kuznetsov (SOVIET UNION)
Nikolai Dmitriyevich Kuznetsov was Klimov's deputy throughout World War 2, and managed GAZ-16 at Ufa. In 1948 he was promoted to General Constructor, setting up his KB at GAZ-19, Kuibyshyev. He clearly had a large engineering team because work began almost simultaneously on 4 extremely challenging turboprop projects, the NK-2, -4, -6 and -12; the staff included over 240 German prisoners taken from wartime gas-turbine development teams. The following are in numerical, not chronological, order.

The NK-2 was a single-shaft axial turboprop run in early 1950 and qualified in 1954 at 3,730 kW—and later at 3,805 kW—not proceeded with. The NK-4 became the AI-20, described under Ivchyenko (*qv*). The NK-6 was a developed NK-2 with reduced weight and improved economy, flown in a Tu-4 in about January 1953 and considered for re-engining that aircraft as the TV-2 at 3,760 kW; after much effort the 4,476-kW TV-2F was cleared to power the Tu-91 with long extension shafts driving an 8-blade contraprop, but the aircraft was not built in series.

The giant NK-12 was an incredible technical achievement which could have been a disaster but instead is still in production. Created by an almost wholly German team, led by Dipl-Ing Ferdinand Brandner (an Austrian), it was a single-shaft turbo-prop which is still more powerful than any successor. The 14-stage axial compressor was designed to handle 62 (later 65) kg/s at a constant speed of 8,250 (later 8,300) rpm, the pressure ratio being varied by several blow-off valves between 9 and 13. The cannular combustor contains 12 conical flame tubes, delivering into a 5-stage turbine. The gearbox provides two co-axial shafts for the AV-60N contra-prop with 8 solid blades of 5.6-m diameter, each half-propeller being able to be pushed round on the ground without moving the other. Cruising propeller rpm is 750, in extremely coarse pitch, with electric (today electronic) control of fuel flow to maintain sea-level power to 8 km and 66 per cent power to 11 km. The prototype ran in 1951, first flight was in a Tu-

Above left *Kuznetsov's mighty NK-12MV is still being manufactured in small numbers more than 35 years from the start.*

Left *The Kuznetsov NK-8-2 commercial turbofan is seen here with reverser, which, like the engine, is based on Rolls-Royce concepts of the 1950s.*

Below *This view of the Kuznetsov NK-8-4 explains the fact that there are twice as many fixed stator vanes ahead of the bypass flow as there are in front of the core. In this engine there is only a small speed difference between the LP and HP spools.*

4LL in 1953 (this aircraft crashed) and production was authorized as the NK-12 in January 1955 at 8,000 kW, followed by the NK-12M of 8,948 kW, the 12MV (1959) at 11,033 kW and the 12MA at 11,185 kW. A typical engine length is 6 m and dry weight 2,350 kg.

In about 1960 the NK bureau began work on the NK-8 turbofan. This was not ready for the prototype Il-62, but was finally qualified (behind schedule) in 1966 complete with reverser. Features include an inlet with 15 fixed stators (30 ahead of the fan bypass flow), a 2-stage fan with swept anti-flutter blades (pr 2.15 at 5,350 rpm) rotating with two IP compressor stages, a 6-stage HP spool largely of titanium with overall pr 10.8 at HP rpm of 6,950, annular combustor with 139 burners, single-stage HP and 2-stage LP turbines all with shrouded blades, and in some applications a Rolls-type reverser. Take-off ratings are 10,500 kg for the NK-8-4 and NK-8-2U, and 9,500 kg for the NK-8-2. A typical weight is 2,400 kg. The NK-86 is a developed version rated at 13 tonnes, which powers the Il-86. A related engine is the NK-144, power unit of the Tu-144 (not 144D), with 3-stage IP, 11-stage HP, pr (static) of 15 and airflow of 250 kg/s. It ran before January 1964 and take-off (afterburner) rating grew to 20 tonnes, for engine weight of 2,850 kg. A similar engine powered the original Tu-22M (Backfire) but this family of engines was conspicuously absent from a recent Soviet eulogy of Kuznetsov's work.

L

Lawrance (USA)

Charles R. Lawrance designed racing-car engines from 1910, but in 1917 founded the Lawrance Aero-Engine Corporation to create a flat-twin air-cooled engine for the Shinnecock lightplane. The firm had one room on New York's Broadway, buying parts from outside. An amazing feature of the 28-hp Model A was that both pistons worked on the same crank! Despite the resulting vibration, and a weight near 200 lb, 450 were built under licence by Excelsior for 'Penguins' (Army non-flying trainers). The Navy became interested and got Lawrance to develop the Model N, giving 40 hp for a weight of 80 lb, and by 1918 this became the 3-cylinder Model L, of 65 hp

and weighing 147 lb, sold in useful (20 to 30) numbers to the Army also. Lawrance was thus able to move into a small loft building and do his own experimental work, and in 1919 began developing two 9-cylinder radials, the J for the Navy and R for the Army.

The R was almost three Model Ls, with the same 4.25 × 5.25-in cylinders and with a carburettor for each group of three, capacity being 670 cu in. A 50-h test was passed in 1921 at 147 hp at 1,600 rpm, weight being 410 lb. The Model J had cylinders 4.5 × 5.5 in, 787 cu in, and the J-1 gave 200 hp at 1,800 rpm, for a weight of 476 lb. These were the first air-cooled radials in production in the USA, and it was a major act of faith by the Navy in an untried engine.

After thinking it could put its tiny competitor out of business, mighty Wright, in the person of F.B. Rentschler, bought up Lawrance, but retained him as vice-president along with several engineers. The result was the famed Whirlwind (see Wright). After Rentschler resigned to form Pratt & Whitney Aircraft, Lawrance became president of Wright.

Leduc (FRANCE)

René Leduc was the greatest pioneer of the ramjet aeroplane. In 1935 years of research led to a small unit which developed 4 kg thrust at 300 m/s (671 mph). He immediately began design of a remarkable piloted aircraft, the 0.10, with an integral ramjet fuselage internally divided into concentric zones, each with a ring of burners, the maximum thrust at 900 km/h at sea level being 2,250 kg. Construction began in 1938 and was finished in 1946. Gliding flights from a Languedoc were followed on 21 April 1949 by the first flight of a true ramjet aircraft, over 800 km/h being attained in a climb at high altitude with only half the burners working. Two more aircraft followed, and the fourth was the enlarged and much more advanced 0.21, with a design thrust at 1,000 km/h of 6,500 kg. First flight was on 16 May 1953. Last came two impressive 0.22s, with an integral Atar turbojet for take-off and planned to reach Mach 2, thrust at sea level then being 60 tonnes. The first flew on the Atar alone in December 1956, but government support tapered off and so did Leduc's dreams.

Le Rhône (FRANCE)

The Soc des Moteurs Le Rhône produced its first rotary in 1910, and between then and the takeover by Gnome in 1914 established its products as being in many ways superior to the Gnome. Its two design engineers were wisely permitted to continue development during the war, and production of 9-cylinder Le Rhônes was on a large scale, with different cylinder sizes giving 80, 110 or 130 or (not built in numbers) 180 hp.

Thulin built them in Sweden, Union Switch &

Rear view of a Le Rhône 9C, showing the extensible induction pipes, twin cylinder-head valves worked by a single rod and dual plugs. Unlike many rotaries the Le Rhône could idle smoothly at 600 rpm.

Signal in the USA, the German Oberursel designs were often more Le Rhône than Gnome, and in the Soviet Union it was the M-2. All Le Rhônes had a cylinder with a cast-iron liner fed with mixture via prominent copper pipes. These had a telescopic section to accommodate changes in cylinder length, it being possible to adjust compression by varying the number of turns the cylinder was screwed into the crankcase. At the head were inlet and exhaust valves both operated by a single push/pull rod; thus, the timing diagram was quite unlike that of a Gnome.

Oddest of all, the conrod big ends were in the form of curved shoes, three short rods, three slightly longer and three longer still. They drove on three concentric bronze-lined circular grooves on the inner faces of two steel discs linked together facing each other. Each rod occupied almost 120° of its own particular groove. On the outer face of the two discs was a flange locating a ball race running on the stationary crankpin. Le Rhône were justified in claiming lower fuel and oil consumption than, say, a Gnome Mono. The '80-hp' Le Rhône with cylinders 105 × 140 mm, actually delivered 93 hp at 1,200 rpm for consumption of 6-7 gal/h, compared with 10 for the 100-hp Mono; its oil consumption was half, at 1 gal/h. It was also cheaper, and weighed 240 lb (UK production) compared with 300 lb for the Gnome.

Liberty (USA)

This famous engine was designed by Jesse Vincent of Packard and E.J. Hall of Hall-Scott, in a Washington hotel suite between 30 May and 4 June 1917. The first engine, a V-8, was on test by 3 July! Altitude testing was done on 14,109-ft Pike's Peak, Colorado. The Liberty was a massive US War Department project to provide vast numbers of aero engines quickly. It was inevitable that the configuration chosen should have been a separate-cylinder water-cooled V, and the cylinder was virtually the same as the 5 × 7 in Hall-Scott; other features were incorporated from a new V-12 by Packard. The angle between the cylinder banks was only 45° to reduce width, and another unusual feature was the coil ignition. The overhead valve gear was exposed, and along each side of the crankcase was a heavily-braced mounting platform. The intention was to produce Liberties in 4-, 6-, 8- and 12-cylinder versions, the first engine being a Liberty 8, rated at 250 hp. Only very few Liberties were made except for the 12, and this quickly went into mass-production at established auto factories, 20,478 being delivered, almost all of them prior to November 1918.

Typically weighing 790 lb, this V-12, 1,649-cu in (same as the Merlin) engine was a sound and generally reliable powerplant, though quality apparently varied depending on the builder. Many thousands continued in military and civil use until 1933 (RAF) and 1934 (US Army), experimental versions having geared drive, 2-speed geared drive, mechanically driven supercharger, turbosupercharger, cast blocks (eliminating the persistent cracking of the

No engine in aviation history can quite equal the Liberty 12's record of quick design, quick qualification and quick mass-production. Remarkably, it also had a very long active life. This particular engine is in the Science Museum, London.

welded water jackets caused mainly by detonation on poor fuel) and, by Allison, a complete inversion of the engine with air-cooled cylinders.

Lockheed (USA)

It is not widely known that this company was a pioneer of the turbojet. Nathan C. Price, a steam-turbine engineer, flew a Travel Air biplane on a steam turbine on 12 April 1933. It was intended only as a publicity stunt to draw attention to the turbine, but was amazingly quiet, smooth and successful (the Boeing School took up the idea). By 1940 Price was with Lockheed, and by this time was firmly set on jet propulsion, using a steam turbine. In 1940 he switched to the L-1000, an incredible engine for which the L-133 canard jet fighter was designed, with integrated propulsion and boundary-layer control, and with reaction-jet controls!

The L-1000 was planned to be more efficient than piston engines, with axial and reciprocating (later axial plus axial) compressors in series, an annular combustor and two intercoolers downstream of the compressors. Sea level thrust was to be 5,500 lb. Hall L. Hibbard rightly thought that no engine firm would want to be involved, but it should have been obvious that the L-1000 was a giant task. In 1943 long-term support was obtained, with designation XJ37, but in October 1945 Lockheed wisely gave the project to Menasco to carry on, under licence. Prototype XJ37s were run, but Menasco in turn handed the project to Wright, which terminated this remarkable programme in 1952.

Lorraine (FRANCE)

The Soc Nationale de Construction de Moteurs (Lorraine-Dietrich) began designing water-cooled aero engines in 1915. At least two in-line prototypes were tested to perfect the Mercedes-style separate-cylinder design before production of the V-8 type began in August 1917, mainly in the form of the 275-hp 8B and 8Bd. There followed a succession of V-12s and 'broad arrow' W-12s, the latter configuration producing the 12Ed of 1922 with cylinders 120 × 180 mm (24.4 litres), rated at 450 hp and used for several famous long-distance flights. This and related engines had cylinders in pairs sharing a common welded water jacket, with overhead camshafts and exposed valve gear. By this time Lorraine had overcome the problem of persistent crankshaft breakage, thought to have been caused by poor mixture distribution.

By 1926 the company had switched to cast-block engines, with steel sleeves screwed into the light-alloy block and in most cases with a mix of plain and roller main bearings, 4 valves per cylinder driven by twin overhead camshafts and, from 1928, a supercharger

A mécanicien, possibly of Air Union, photographed at Croydon in about 1923 with a Lorraine-Dietrich V12 of 400 hp.

and geared drive. This family was named after birds. The Eider of 1928 was a massive 45-litre V-12 with 170 × 165 mm cylinders, rated at 1,050 hp. The Courlis of 1929 was a W-12 with 145 × 160 mm cylinders (32.1 litres), rated at 600 hp. The Petrel of 1932 was a V-12 with 145 × 145 mm cylinders (29.5 litres), rated at 500 hp, but developed to 860 hp in the Petrel 12Hars of 1938 which drove a contraprop in the FK.55 fighter. Last came the Sterna of 1937, a V-12 with 30 litres and type-tested at 810 hp at 2,575 rpm, with 1,200 hp in prospect—in each case at 4 km (13,120 ft). Lorraine built small radials, such as the 120-125-hp P5 of 1933, but poor sales resulted in bankruptcy in 1940.

Lotarev (SOVIET UNION)

Vladimir Lotarev is the General Constructor in charge of the KB at Zaporozhye previously directed by A.G. Ivchyenko. His obviously large team have produced some of the most advanced gas turbines in the Soviet Union, three of them being in production.

First came the smaller of two high-bypass turbofans, the D-36, developed for the Yak-42. This has three shafts, with a titanium fan, LP spool with variable inlet vanes and steel HP spool, giving overall pr of 20, bypass ratio being 5.6. The 28-burner annular combustor is fabricated with integral HP turbine entry vanes, inlet gas temperature being 1,177°C. The air-cooled HP turbine and the IP both have one stage, the LP having two. Normal take-off rating is 6.5t, with sfc of 0.36. Dry weight is 1,100 kg. The D-36 ran in 1973 and also powers the An-72 and 74.

Next came the world's most powerful helicopter engine, the 11,400-shp D-136 for the Mi-26. This has a 6-stage LP spool, 6-stage HP, combustor scaled down from that of the D-36, single-stage HP turbine (1,205°C), single-stage LP and 2-stage free power turbine. Construction is modular, take-off sfc 0.4365, and dry weight 1,050 kg.

Latest Lotarev engine known is the D-18T, the larger high-ratio turbofan which may have been started at the same time as the D-36 but was not required until the giant An-124 was ready in 1983-84. The bypass ratio is 5.8, the titanium fan has 33 blades, overall pr is 27.5, the combustor is a scaled-up edition of that of the D-36, and the number of turbine stages is 1-1-2, maximum gas temperature being well up to the best Western practice at 1,327°C. The D-18T has an integral fan reverser similar to that of the CF6-50, and accessories are grouped round the fan case. Take-off rating is 23,430 kg, with sfc of 0.36; dry weight is 4,100 kg.

Lycoming (USA)

In numerical terms this is currently the world's No 1 producer of aero engines. Founded in Williamsport, Pennsylvania, in 1908, it made car engines from 1910, by 1924 powering 57 *makes* of automobile. In 1929, as a subsidiary of E.L. Cord's Auburn Auto Company, it produced its first aero engine, the R-680 radial with 9 cylinders 4.625 × 4.5 in, each with two valves, and a conservative design which soon produced outstanding reliability, at ratings from 200 to 285 hp. In 1932 Lycoming Manufacturing became part of the conglomerate Aviation Corporation (from 1947 renamed Avco). Production of the R-680 continued until after World War 2, to a total exceeding 26,000. In 1938 a new line of light-plane engines had been designed by Harold Morehouse starting with the O-145 flat-4. The left and right pair of 3.625 × 3.5-in cylinders was cast integral with half the mild-steel crankcase, the cast aluminium heads being held by studs. Opposite cylinders were staggered so that the crankshaft could have 4 throws. The O-145 started at 50 hp, weighing 152 lb, and at termination in 1950 had progressed to 75 hp in both direct-drive and geared versions. In 1939 Lycoming designed the O-235, with 4 cylinders 4.375 × 3.875 in, starting life at 104 hp, and the O-290 of 125 hp with bore increased to 4.875 in. This larger cylinder was also soon used in the flat-6 O-435 family of 190-220 hp, and all these engines had an aluminium-alloy crank-

Top *Lotarev's D-36 was the first 3-shaft turbofan to be developed for service other than by Rolls-Royce. It powers the Yak-42 and has a good record.*

Above *The D-136 turboshaft engine, rated at 11,400 shp, makes possible the big Mi-26 helicopter. The West has no engine in this class.*

Below *In May 1985 the giant An-124 visited Paris, powered by four D-18T turbofans. In this aircraft reversers were fitted.*

case split on the vertical centreline to which the steel cylinder barrels were attached by studs. By 1950 this range had been joined by the 8-cylinder O-580 family of 320-400 hp, originally considered for buried wing installations. All these engines had many common parts.

The flat-8 was almost a throwback to a major Army programme begun in 1935 to find an engine to fit inside wings. After a year testing a liquid-cooled 'hyper' cylinder, the O-1230 flat-12 was placed under contract and developed with speed and assurance, largely because Lycoming put more of its own money into the project. The 1,234 cu in engine ran in 1937, and flew in 1938 powering the XA-19A in a conventional nose installation. It was qualified at 1,000 hp in 1939, but more power was now needed, so Lycoming went ahead with the H-2470, a twinned O-1230. Navy funds supported it, and the first run in July 1940 was encouraging, so a production order was placed (for the F14C-1) in May 1942. In the event the only aircraft powered by this 2,200-hp liquid-cooled unit was the Convair XP-54.

Top left *Lycoming's R-680 established the firm in the aero-engine business. This 300-hp wartime E3A model weighed 515 lb.*

Top right *Harold Morehouse, previously with Continental, boldly designed Lycoming's first opposed engine, the O-145, with a monobloc casting containing the crankcase and cylinders.*

Above left *By late 1939 Lycoming's O-1230 flat-12 was giving 1,200 hp for a weight of 1,325 lb, and with a total height of 37 in. The Army wanted even more than this.*

Above *First run in July 1940, the Lycoming XH-2470 was two O-1230s superimposed. A special plant at Toledo, Ohio, was to build it for the F14C, the first liquid-cooled Navy fighter since 1925, but this potential 3,000-hp engine was cancelled in 1943 on grounds of too-late timing.*

Far more remarkably, in 1941 an Army contract launched the XR-7755, the biggest one-unit piston aero engine and one of the heaviest aero engines of all time (7,050 lb in contraprop form). With 9 banks of 4 whopping 6.375 × 6.75 in cylinders (7,755 cu in) it

Left *Lycoming's XR-7755 remains a mystery; what aircraft could have used this colossus among piston engines?*

Below *Today Williamsport builds them a bit smaller, an example of the world's biggest-selling range being this TIO-540, which weighs 450 lb and gives 250 hp up to 15,000 ft; other versions sustain up to 360 hp to this height, all cylinders exhausting through the single turbo.*

Bottom *A conceptual mock-up of the 200-hp size, smallest of the new Avco Lycoming RC (rotating combustion) aero engines. Developed in partnership with John Deere, these stratified-charge engines could oust the reciprocating engine by the year 2000.*

Avco Lycoming Williamsport's LT101 range of turbines in the 600-hp class show how today the engine is dwarfed by the inlet, gearcase and accessories. This is a digitally controlled LTS101-750A-3.

gave 5,000 hp in 1944 and aimed at 7,000 hp, but no application was ever announced.

Today Avco's aero engines are produced at Stratford, Connecticut (gas turbines), and Williamsport (general-aviation engines including small gas turbines). The company's best-selling piston engines comprise three flat-4s, the O-235 (4.375 × 3.875 in), O-320 (bore increased to 5.125 in) and O-360 (5.125 × 4.375 in), with ratings of 115-282 hp, two families of flat-6s with O-360 cylinders, the O-540 and the redesigned O-541 range, and the corresponding flat-8, the O-720. Almost all have direct injection and many are turbocharged. Piston-engine deliveries from Williamsport exceed 250,000.

For nearly a decade Williamsport has watched the growth of previously absent competition, notably from derivatives of air-cooled and water-cooled car engines. A vice-president for new product development was appointed, and at last in 1985 a major new line of business was announced. In 1984 the last foothold that the once mighty Curtiss-Wright company had in complete aero engines was purchased by John Deere, the giant of agricultural machinery. The vast Wood-Ridge plant had from 1958 been developing Wankel-type RC (rotary combustion) engines, as also had NASA at Lewis Research Center. Deere intends to take RC engines further, for many applications, and in March 1985 announced an agreement with Avco Lycoming under which Avco will pay a share of the development, plus a

royalty on each engine, and in return have world rights to resulting aviation engines. Avco has started with a twin-rotor liquid-cooled engine of 350-400 hp, lighter than ordinary piston engines and with stratified-charge combustion adaptable to almost any common fuel. Flight test is due to start in 1987 and deliveries are expected in 1990. It looks as if Wankel's smooth-running engine will now penetrate the aviation market at last, after 30 years.

In 1951 Avco Lycoming hired Dr Anselm Franz, lead designer of the Jumo 004B in World War 2. A year later he began work on the T53 turboshaft for Army helicopters, the prototype running in 1953 at 600 hp. Features included a 5-stage axial and 1-stage centrifugal compressor, folded annular combustor, and single-stage gas-generator and free power turbines. Since then over 19,000 T53s have logged over 36 million hours in turboshaft and turboprop forms at powers up to 1,550 shp, some licence-made by KHD, Piaggio, Kawasaki or AIDC-Taiwan. In 1954 work began on the larger T55, with airflow raised from 10 to 20 lb/s (later models, from 12.2 to 27 lb/s). Some 5 million hours have been flown by 3,600 T55 engines, almost all in Chinooks. After building a few PLF1 and related engines, Lycoming Stratford's wish to launch a turbofan was realized in 1969 in the

A cutaway showing the geared drive to the fan (and supercharging LP stage) of the Avco Lycoming ALF 502R which powers the BAe 146. The axial/centrifugal core and reverse-flow combustor are taken from the T55 which powers the Chinook.

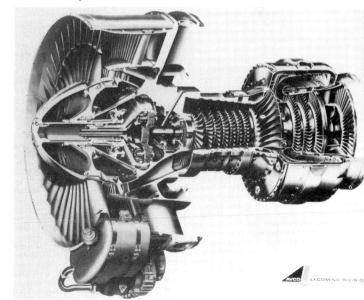

ALF 502, using the T55 core plus a reduction gear and fan/accessory module. Two versions are in production: the 502L family (Challenger) have bypass ratio 5, pr 13.6 and 7,500 lb rating, while the 502R (BAe 146) has figures 5.71, 11.6 and 6,700 lb, and at 1,270 lb is slightly lighter. Stratford also developed a family of simple, robust, modular single-shaft engines in the 600-hp class known as LTS 101 (turboshaft) or LTP 101 (turboprop), military designation being T702. These have been assigned to Williamsport for production. With Army funding Stratford has developed the PLT 34 in the 1,200-hp class with technology intended for the next century.

Lyul'ka (SOVIET UNION)

Arkhip Mikhailovich Lyul'ka (the apostrophe replaces a Russian character that separates the syllables) was one of the greatest pioneers of the axial turbojet. His 1936 calculations at the Kharkov Aviation Institute met with scepticism, but, aided by famed designers Kozlov and Shyevchyenko, he was permitted in 1938 to begin work on the VRD-1 turbojet in Leningrad. This 8-stage engine was to develop 600 kg thrust (a little more than its weight) at 700°C, but had to be shelved in 1941. In late 1942 Lyul'ka was able to return to the beleaguered city where he tested centrifugal and axial compressors, ran the VRD-2 axial at 700 kg in 1943 and late in that year headed a team that began design of the VRD-3, later called S-18 and finally TR-1.

This much better 8-stage engine had an annular chamber, single-stage turbine and many new features; it gave 1,300 kg at 6,950 rpm in 1944 for a weight of 885 kg. Four powered the Il-22 four-jet bomber first flown 24 July 1947. The Su-10 had four TR-1As each rated at 1,500 kg. In 1946 work began on the VRD-5 (TR-3) which in 1950 achieved its design rating of 4,500 kg, subsequently being built in small series at 4,600 kg for various prototypes; it had a 7-stage compressor and automatic single-lever control.

From 1950 Lyul'ka was a General Constructor, the TR-3 being restyled AL-5. It later developed 5,500 kg for such types as the Tu-110. So far Lyul'ka had been thwarted by failure of 9 aircraft types to progress beyond the prototype stage, but his next engine, the TR-7 (AL-7) scored a major success and remained in production from 1954 until after 1970. The first large turbojet to have supersonic flow through the first 2 stages of its 9-stage (originally 8) compressor, this impressive engine has an annular chamber, 2-stage turbine overhung behind the rear bearing and, in most applications, a particularly good afterburner giving some 40 per cent augmentation. First run was in 1952, and the main version is the AL-7F-1 rated at 7,000 kg dry and 10t with afterburner. The AL-7PB is

a marinized 6,500 kg engine for flying boats, later redesignated AL-7RV. From this important engine Lyul'ka developed the AL-21, installationally interchangeable but with considerably greater airflow and numerous other improvements. Large numbers continued in production until 1984, mostly in AL-21F-3 form rated at 7,800 kg dry and 11,200 kg with maximum afterburner. A visible efflux at high power is usual, and this has led some observers to consider that a vectored twin-nozzle version of the unaugmented engine powers the Yak-36MP. Lyul'ka almost certainly has other production engines used in new combat aircraft.

Manly (USA)

On 1 June 1898 Charles Matthews Manly reported to the august Professor S.P. Langley of the Smithsonian Institution in Washington DC without even waiting to graduate from Cornell. From then until December 1903 he was engine designer and pilot of Langley's 'Aerodromes'. His engine was derived from the Balzer auto engine, but considerably redesigned. A static radial, it had 5 cylinders each 5 × 5.5 in (540.2 cu in), with closed-head steel barrels only 0.0625 in thick with shrunk-in cast-iron liners of the same thinness, the forged head and 0.02 in water jacket being brazed on. The master rod ended in a sleeve encircling the crankpin and fitted with a bronze liner, each split at 90° to the others to permit assembly. The outer periphery of the sleeve was truly cylindrical (except at the junction with the rod) and the other four rods all bore on this sleeve via bronze shoes. Pistons were cast iron, and a side extension to the combustion chamber contained an automatic sprung inlet valve opposite a push-rod-driven exhaust valve. This remarkable engine weighed 207.5 lb including water and gave 52.4 hp at 950 rpm, the weight/power ratio of 3.96 not being equalled until well into the First World War.

Maxim (UNITED KINGDOM)

Sir Hiram Maxim, wealthy expatriate American, has been described as conceited, pigheaded and much more besides, and he certainly was one of the many

Sir Hiram Maxim is not seated but is actually lifting one of the remarkable 180 hp compound steam engines which powered his giant 1894 biplane.

'chauffeur-minded' aviators who disregarded the problem of how to learn to be a pilot; but his giant biplane of 1894 had remarkable engines. Left and right steam engines drove handed (opposite-rotation) propellers, and it was all on a vast scale. High-pressure boilers burning naphtha fed the two compound engines, each with one HP (320 lb/sq in) and one LP cylinder, each engine weighing only 310 lb but putting out 180 hp (Maxim was sure 250 hp was ultimately attainable). Boilers, etc, added 1,800 lb, plus 600 lb of water. Maxim finally concluded that the internal-combustion engine was better.

Menasco (USA)

Al Menasco established Menasco Manufacturing Corporation in Los Angeles in 1926, the first product being an air-cooled rebuild of a wartime Salmson radial. In 1929 he launched the 4-A inverted 4-in-line with 4.5 × 5.125 in cylinders (326 cu in), with exposed valve gear operated from a camshaft in the crankcase. This engine was rated at 90 hp, but Menasco was an enthusiast for racing and by 1931 had brought out the 95-hp B-4, 125-150-hp C-4 (with bore increased to 4.75 in) and B-6 with 6 original-bore cylinders. Usually the 4s were named 'Pirate' and the

6es 'Buccaneer', but as the 1930s progressed there were numerous hot-rod versions. The ultimate baseline engine was the 6CS Super Buccaneer with large-bore cylinders (545 cu in), normally a 260-hp engine but in racing trim giving well over 300 hp. In 1938 the M-50 flat-4 was launched but it was a marketing error, and the last news of the firm was its brief encounter with the complicated Lockheed XJ37 turbojet.

Menasco's Unitwin comprised two supercharged Buccaneers geared to a single propeller shaft. In an alternative scheme two Menascos were fitted flat, driving a pair of propellers side-by-side in the nose of the Alcor DUO-6.

Having in 1912 pioneered the welded-steel cylinder that later formed the basis of Rolls-Royce and Liberty engines, the Mercedes (Daimler Motorenwerke) was by a narrow margin the No 1 German wartime engine. This is a Mercedes D III (D for Daimler) of 1915, with 140 × 160 mm cylinders rated at 160 hp.

Mercedes (GERMANY)

Strictly the name should have French accents because it was that of a wealthy Frenchman's daughter. It was originally a brand name of the famed Daimler company of Stuttgart, which built its first aero engine in 1910, a water-cooled 4-in-line. Subsequently Daimler (Mercedes) produced an inverted 4 and a geared 8-in-line, but more than 99 per cent of its wartime output was of 6-in-lines.

The Mercedes 6-in-line of 1913 was at least as important as the Austro-Daimler in establishing such engines as almost standard for aircraft of the Central Powers, and they certainly were the first to make a success of the steel cylinder with welded water jackets. Like the Austro-Daimler there were twin carburettors and twin magnetos, but the valves were driven by an overhead camshaft which usually had a lever at the rear with which it could be moved axially to bring into action half-compression cams to facilitate starting. Almost all models had the cylinder barrel and head machined from steel forgings, screwed together and with valve pockets, guides and ports welded on. Jackets were three or four thin steel pressings welded together on each cylinder. Most cylinders comprised a cast-iron skirt welded (often also screwed) to a domed steel crown. The sump sloped down to the rear, and all engines had scavenge pumps and separate oil tanks. The 1913 engine weighed about

209 kg and gave 100 hp at 1,300 rpm, with 14.78 litres from 140 × 160 mm cylinders. During the war this engine was increased in power in stages to 185 hp (Type IIIb), many thousands being made. The other mass-produced series (Type IVa) had 160 × 180 mm cylinders (21.7 litres) and was rated at 260 hp, usually for large bombers. Aero work ceased in 1919-26, subsequently being resumed as Daimler-Benz (*qv*).

Metrovick (UNITED KINGDOM)

The Metropolitan-Vickers Electrical Company, of Trafford Park, Manchester, had a background of axial steam turbines when in August 1937 it was invited to build experimental gas-turbine parts by the Royal Aircraft Establishment. A.A. Griffith and Hayne Constant had pioneered axial compressors and turbines at the RAE from 1926, but real progress was made only after 1937 when—thanks to Whittle—the Air Ministry had at last begun to take gas turbines seriously. GEC and Parsons also received contracts for hardware, but Metrovick was the only non-aero firm to go on and produce complete engines. Its B.10 compressor first ran in December 1939, the drive turbine in May 1940 and the complete (clumsy) turbo-jet in October 1940. Then came the D.11 with straight-through flow through tandem axial spools totalling 17 stages, but efficiency began to fall at pr above 2.02.

In July 1940 Metrovick put together a full design team under Dr D.M. Smith, with major roles taken by Dr I.S. Shannon and K. Baumann and helped by Constant. In July 1940 this team began development

First British axial turbojet, the Metrovick F.2 began at 1,750 lb thrust with a 9-stage compressor and 2-stage turbine and by 1946 gave 3,500 lb with a 10-stage compressor and 1-stage turbine!

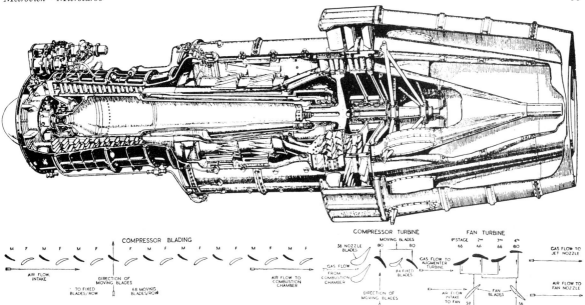

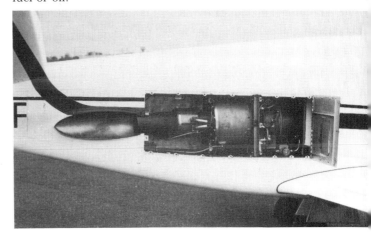

Above *The Metropolitan-Vickers F.3 aft fan was a pioneer engine based on Whittle concepts and Dr Smith's prowess with axial machinery. Note the interlinked turbine and fan stages. Copyright Iliffe & Sons Flight.*

Below right *The little Microjet 200 is powered by two Microturbo TRS 18 turbojets each of 247 lb thrust.*

of the F.2 axial turbojet. Blading for the 9-stage compressor was designed at RAE; in 1941 this far outperformed the D.11 and in 1942 it maintained adiabatic efficiency of 88 per cent to pr of 3.2. The first F.2 ran in December 1941, and was the first non-German axial in the world. After many changes the third engine flew in the tail of Lancaster *LL735* on 29 June 1943. On 13 November 1943 two F.2s powered F.9/40 (Meteor) *DG204* on the first flight outside Germany of an axial-engined aircraft. Rating was then 1,800 lb, but after three redesigns the F.2/4 was built in small series in 1945 with a 10-stage compressor and thrusts of 3,500, 3,750 and 4,000 lb, weight being around 1,800 lb. The engine was named Beryl, first of a 'precious stone' series. Metrovick achieved promising results from August 1943 with the F.3, an F.2 with a thrust-augmenting ducted fan which boosted output from 2,400 to 4,600 lb. In 1945 the F.5 was run at 4,830 lb with a 2-stage UDF (unducted fan), today reinvented by GE! In 1946 Smith began his masterpiece, the big F.9 Sapphire, with a disc-built compressor that was by a clear margin the best in the world at the time—the engine ran in April 1948. Perhaps shortsightedly the firm had decided in 1947, under Ministry pressure, to get out of aviation; the Sapphire was handed to Armstrong Siddeley.

Microturbo (FRANCE)

France appeared to have a near-corner in the market for small turbine engines with Turboméca, but that company's engines are giants compared with Microturbo! Formed at Toulouse in 1960 to build tiny gas-turbine starters, this energetic firm has sold small gas turbines for almost every purpose. The TRS 18-046 was designed to give self-launch capability to gliders, but is also used in many twin-engined high-speed applications. A modular and very simple centrifugal engine, it weighs 37 kg, is only 650 mm long without jetpipe, and has a continuous rating of 102 kg at 45,000 rpm. Start sequence (on ground or in flight) is automatic, and it can use almost any aviation fuel or oil.

A. S. Mikulin became famous for his big and reliable V-12 liquid-cooled engines. This model, the AM-38F of 1,700 hp, was one of the engines fitted to the Il-2 Stormovik, of which 36,163 were delivered.

Mikulin (SOVIET UNION)

Aleksandr Aleksandrovich Mikulin was one of the first engineers to work at NAMI (scientific auto motor institute), having been engaged in engine design since 1916. In 1925 he used Junkers techniques in the AM-13, a large water-cooled V-12 qualified in 1928 at 880 hp at 2,150 rpm, and he collaborated with N.R. Brilling on the Soviet derivatives of the Jupiter, notably the 2-row 18-cylinder M-18 which gave severe problems with the complex valve gear. More important was Mikulin's overseeing the licensed BMW VI and its development into the M-17 family.

In parallel Mikulin studied available hardware and in 1930 obtained permission to design the AM-30 in an attempt to create the best V-12 possible. It used modified BMW VI cylinder blocks (160 × 190 mm cylinders, 46.7 litres), an HS12 rear wheelcase, Allison supercharger and RR Buzzard reduction gear. In 1931 it was qualified at 660 hp at 2,000 rpm. It was the starting point for a major series of large engines, the first production type being the AM-34 built in at least 14 versions between 1932-39 with compression ratio 6.25, 6.6 or 7.0 and with powers 690 hp (34) to 930 hp (34R, F), 900 hp (FRN), 950 or 970 hp (R/RN) cr 7, 1,200 hp (FRNV) or 1,275 hp (RNF).

The AM-35 had a new cylinder head and improved supercharger, qualified in 1939 at 1,200 hp, the AM-35A reaching 1,350 hp. The AM-37 of 1940 reached 1,380 hp, or 1,400 as the 37F. Vast numbers were made of the AM-38 qualified in 1941 at 1,550 hp, later 1,665 hp, and the 38F of 1,700, 1,720 and 1,760 hp. In 1942 the AM-39 was qualified at 1,870 hp,

followed by the FN-2 at 1,850 hp, the A at 1,900 and FB at 1,800. The family ended with the AM-42 of 2,000 hp, AM-43 of 1,950 to 2,200 hp, and AM-47 of 1946 of 2,700 to 3,100 hp (AM-47F).

Mikulin ended the war with great power, and opened a giant turbojet KB with Tumanskii as his deputy. Work began on the AM-2, -3, -5 and -9. The AM-2 was tested 1953 at 4,600 kg but was not adopted. The AM-3 was a large but simple engine developed by P.F. Zubets' team, with an 8-stage compressor (135 kg/s, pr 6.4 at 6,500 rpm), thrust being 6,750 kg in 1952, 8,200 kg in AM-3M in 1954, 8,700 kg in AM-3M-200 and 9,500 kg in 3M-500; service designation is RD-3 and internal KB designation M-209. Many hundreds remain in use. The slim AM-5 was rated at 2,700 kg (3,040 with afterburner) but was not adopted. It was replaced by the improved M-205, or AM-9 (TsIAM designation), with 9-stage compressor (pr 6.3 rising in later versions to 7.14), annular combustor and 2-stage turbine, diameter being 813 mm and thrust of production versions typically 3,300 kg with afterburner. This was picked for the Yak-25 and with afterburner for the MiG-19, but there was upheaval in the KB. Someone had 'informed' on Khrunichyev (aviation minister) and Stalin appointed a commission to investigate him.

Mikulin did all he could do to incriminate the minister, who was arrested and might have been executed. Then Stalin died, Khrunichyev was set free and quickly settled the score by getting Mikulin blacklisted. The engines were all redesignated with RD numbers only, and the KB was led by Tumanskii from 1956 onwards.

Mitsubishi (JAPAN)

This vast company was the first series producer of aero engines in Japan, and in World War 2 it ranked No 1, with 38 per cent of gross national output of engines. Apart from a giant stillborn liquid-cooled unit, all were 2-row radials with 2-valve cylinders and very similar (and often interchangeable) features. In brackets are 1943 Army/Navy joint designations. The 14-cylinder engines were: Ha-6, 825 hp; Ha-26 of 900-950 hp and the related Zuisei and Ha-102 (Ha-31) of 875-1,080 hp; Ha-101 and related Kinsei MK8 and Ha-112 (Ha-33) of 1,280-1,560 hp; and the Kasei MK4 (Ha-32) of 1,460-1,850 hp. The 18-cylinder engines were the Ha-104 of 1,800-2,000 hp, the Ha-211 and related MK9 (Ha-43) of 2,200 hp and the Ha-214 and related MK10 of 2,400 hp. Some of the larger engines were developed in Ru (turbocharged) versions. In 1952 Mitsubishi resumed aero work and has licence-built various engines; it participates in IAE and is developing a small missile turbojet and a large liquid space rocket engine.

Nakajima (JAPAN)

Established in 1914, Nakajima built large numbers of Bristol Jupiter and Lorraine engines in 1927-38 and developed the Ha-1 Kotobuki of 550-785 hp from the Jupiter. In World War 2 it ran a close second to Mitsubishi in engine production, and it was easily 'No 1' in 1944-45. Main engines were: 9-cylinder radials, the Ha-1 already mentioned, Ha-8 and Hikari of 700-840 hp, Ha-20 of 730-820 hp and Ha-26 of 900 hp; 14-cylinder radials, Ha-5 of 950-1,080 hp, Sakae NK1/Ha-25/Ha-102/Ha-105/Ha-115 of 950-1,230 hp, Ha-109/Ha-34 of 1,300-1,500 hp, Ha-41 of 1,260 hp, Mamoru NK7 of 1,800-1,870 hp and Ha-117 of 2,420 hp; 18-cylinder radials, Homare NK9 (Ha-45), Ha-44/NK11 and Ha-145 (all with Sakae-size 130 × 150 mm cylinders, giving 32 litres) of 1,800-2,400 hp, and Ha-217/Ha-46 of 3,000 hp; and the Ha-505 with four 9-cylinder rows and rated at 5,000 hp. After 1945 the company became Fuji, building airframes only.

Napier (UNITED KINGDOM)

David Napier came from Scotland to London in 1808

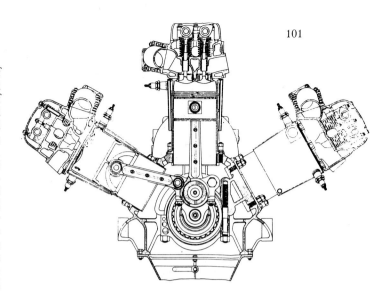

Cross-section of a Napier Lion IIB, a typical Service mark dating from December 1923. It weighed 966 lb and was rated at 470 hp at 2,000 rpm. Cylinders in the three 60° blocks were all aligned, though one has not been sectioned.

to make printing machines. D. Napier & Son Ltd moved to Acton in 1903, became famous for its cars and from 1915 made RAF.3a and Sunbeam Arab engines. In 1916 Montague Napier began working with chief designer A.J. Rowledge on the design, at company expense, of a completely new engine. The result was the Lion, a masterpiece. First run in April 1917, it was a water-cooled W-12, or broad-arrow engine, with 3 banks of four 5.5 × 5.125 in cylinders (1,461 cu in) spaced at 60°. The closely spaced steel cylinders were spigoted and flange-bolted into the aluminium crankcase, with integral combustion heads and steel water jackets welded on. Each row of 4 cylinders was capped by a monobloc aluminium head secured by screwing into it the seats for the 2 inlet and 2 exhaust valves for each cylinder, each head containing the inlet/exhaust ports, water passages and 5 bearings for the twin overhead camshafts. The vertical cylinders had master rods, working on the short and almost unbreakable crankshaft, the other cylinders having auxiliary rods pivoted to the masters by tapered wrist pins. From the start it was a magnificent engine, even the prototype giving a reliable 450 hp at 2,000 rpm. Most service Lions were geared and rated at up to 570 hp at 2,585 rpm, though 2,350 was a common limit. Racing Lions ran at up to 3,900 rpm, giving up to 1,400 hp.

The company's utter complacency sapped the morale of its few designers, and inevitably the Lion became uncompetitive after 1930, though Sea Lions for RAF rescue launches stayed in production to 1943. In 1923 there was a stillborn attempt to get the Lioness, an inverted Lion, into RAF fighters because

Above *The author never flew with a Napier Sabre but will never forget the sound of Typhoon squadrons. As sheer packages of engineering they were impressive. This was the initial production Mk II of 2,400 hp.*

Left *The Napier Nomad passed through two very dissimilar development stages. This was the second, simplified version, with the diesel and gas turbine both geared to the same propeller shaft.*

of the improved pilot view it offered. In 1919 Montague Napier—at that time actually looking beyond the Lion—got Air Ministry backing for an engine of 1,000 hp, the Cub. This had 4 banks of 4 cylinders bigger than the Lion's arranged in the form of an almost flat-bottomed X. Six were built, but the market was not ready.

To try to reverse its decline Napier got Halford in 1928 to design a range of small-cylinder engines of only 404 to 718 cu in capacity. The result was a series of unusual engines of H configuration. These began with the Rapier, with 16 air-cooled cylinders of 3.5 × 3.5 in size (538.8 cu in) arranged in 4 vertical rows driving side-by-side crankshafts, output of production engines being 340 hp. The Javelin inverted 6-in-line of 160-170 hp hardly penetrated the market, but

the Dagger almost saw major production. An enlarged Rapier, with 24 cylinders 3.8125 × 3.75 in (1,027 cu in), it ran at 635 hp in 1934 and was cleared for production in 1936 at 1,000 hp at 4,200 rpm, though for reasons of reliability and maintainability it soon faded from the scene. All these air-cooled engines were fully supported by the Air Ministry, though none was needed.

In 1935 the Air Ministry fell for a proposal that Napier should build a 'Hyper' engine running at high rpm with many small but highly rated cylinders, the output being 2,000 hp. Halford thus created the Sabre, adopting the flat-H configuration with liquid cooling and sleeve valves. The engine comprised upper and lower light-alloy blocks each with 12 opposed cylinders 5 × 4.75 in (2,238 cu in) each with

3 inlet and 2 exhaust ports, and 4-port sleeves based on Bristol practice. At the rear was the 2-speed supercharger and a 4-choke updraught carburettor which in most Sabres was of the injection type. On top was the Coffmann cartridge starter and other accessories. A compact engine distinctive for its thrilling high-rpm sound, the Sabre ran in 1937 and was type-tested at 2,200 hp in 1940, but suffered many prolonged problems. In retrospect the gigantic effort put into it would have been better applied elsewhere, but by 1944 it was becoming reasonably reliable with ratings up to 3,055 hp at 3,850 rpm for a weight of 2,540 lb.

By this time chief engineer, Ernest Chatterton, was busy with the ultimate in economical engines for long-range operation, the Nomad. As first run in 1950 the E.125 Nomad comprised a liquid-cooled flat-12 diesel with big 6 × 7.375 in cylinders (2,508 cu in) with valveless 2-stroke operation, supercharged by an axial and centrifugal compressor in series, with intercooler. The crankshaft drove half a contraprop and an exhaust gas turbine the other half, and for maximum power a reheat chamber and auxiliary turbine were brought in. With much flame and backfiring the power was 3,125 ehp (3,000 shp) for a weight of 4,200 lb. By 1953 the engine had been simplified into the E.145 with the centrifugal compressor, reheat chamber and auxiliary turbine eliminated, drive to a single propeller being by the crankshaft, the gas turbine being linked through a variable-ratio Beier gear. Power went up to 3,135 ehp (with water injection, 4,095 ehp) and weight fell to 3,580 lb. Though sfc set a new low at only 0.345 this engine remained a near miss, and the Shackleton stayed with the Griffon.

Meanwhile Napier had got into gas turbines under A.J. Penn and Bertie Bayne. The 535-hp Nymph was dropped, and so after years of work was the 1,600-ehp Naiad and its coupled version, but the Eland ran on for 9 years (1952-61). Its 10-stage compressor had blades of aluminium-bronze (DTD.197) as did the other Napier gas turbines, and it was on the same shaft as the propeller gearbox, all driven by a 3-stage turbine. Starting at 3,000 ehp (2,690 shp) it grew in

Below *Napier's Double Scorpion was one of many rocket engines that never got into production, though this peroxide/kerosene package did thrust a Canberra to well over 70,000 ft in August 1957.*

Right *Air for the Napier Gazelle was drawn in at 16 lb/s round the edge of the massive inlet casting. At 780 lb the Gazelle did not compare very well with the T58, which, as the Gnome, replaced it in some helicopters.*

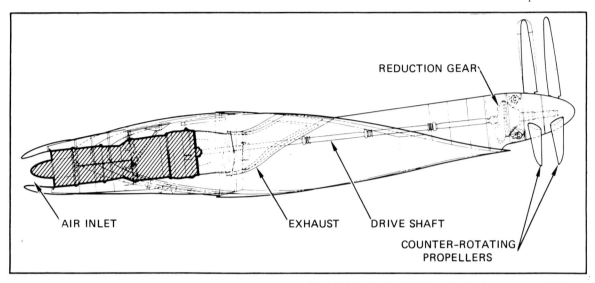

REDUCTION GEAR

AIR INLET   EXHAUST   DRIVE SHAFT

COUNTER-ROTATING
PROPELLERS

Had the Northrop B-35 flying-wing bomber gone into production it would probably have been powered by the Turbodyne in the installation seen here.

1955 to 4,200 ehp with air-cooled blades, and the E.151 (NEl.3) with auxiliary compressor powered the Rotodyne convertiplane. Of the Oryx gas generator for the Percival P.74 tip-drive helicopter the less said the better, but the Gazelle did at last find a real market as an any-attitude free-turbine engine for helicopters. First run in December 1955 at 1,260 shp, it had an 11-stage compressor, 6 flame tubes and 2 + 1 turbine stages, production engines giving up to 1,790 shp. D. Napier & Son also produced ramjets and rockets, but in 1960 was divided, Napier Aero Engines Ltd becoming a subsidiary of Rolls-Royce in 1962 and soon losing its identity.

Northrop (USA)

When John K. Northrop founded his own company in August 1939 his head of research was Vladimir Pavlecka, a Czech, who managed to convince Northrop that in the long term gas turbines would replace piston engines. After much in-house study an Army/Navy contract was obtained in June 1941 for design plus tests of the compressor for an engine of 2,400 hp. In 1942 Pavlecka had a row with the great aerodynamicist Von Kármán and left. His replacement was an Englishman from General Motors, Art Phelan, who pointed out that to drive the compressor needed 7,000 hp and that it made sense to build the whole engine. On 1 July 1943 two engines of 3,800 hp were contracted for, and one of these was the first US turboprop to run complete with propeller in December 1944.

Named Turbodyne, the engine was then scaled up to meet Army needs, and the XT37 was designed in 1945 for 4,000 hp at 35,000 ft for future bombers. Three engines were running in 1947 and achieved

sea-level power rising from 5,150 to 8,000 hp in six months. By 1949 the XT37 with 14-stage compressor (pr 7.5) and 2-stage turbine was giving 10,400 hp at only 815°C, developed by the Northrop-Hendy partnership owned 50-50 with the Joshua Hendy Iron Works. Perhaps shortsightedly the USAF cancelled the EB-35B testbed in October 1949, effectively terminating the project, entirely because Pratt & Whitney had shown that the B-52 could be a jet.

Orenda (CANADA)

In 1945 Hawker Siddeley purchased the wartime Victory Aircraft plant at Malton, Toronto, and formed Avro Canada, which in 1946 took over Turbo Research, a nearby government gas-turbine laboratory. Under chief designer Winnett Boyd the Chinook axial turbojet was designed, and run on 17 March 1948. This had a 9-stage compressor, 6 cans and 1-stage turbine, and gave 2,600 lb thrust, later 3,000. It assisted design of the Orenda, under chief engineers Paul Dilworth and then C.A. Grinyer. The Orenda, which gave its name to the vastly expanded Orenda Engines Ltd in early 1956, was an enlarged Chinook with a 10-stage compressor.

The Orenda 1 ran in February 1949, leading through various marks to the Orenda 9 and 10. These had airflow of 106 lb/s, pr 5.5 and thrust 6,500 lb. The Mk 9 had an alcohol spray fan on the nose bullet and could be installed in a CF-100 in either left or right position; the Mk 10 fitted the Canadair Sabre. The Orenda 11 and 14 were the corresponding versions with a new 2-stage turbine, reducing weight to 2,430 lb and increasing thrust to 7,500 lb at 7,800 rpm.

In 1953 design began on the totally new PS.13 Iroquois to power the CF-105 Arrow. An outstandingly advanced 2-spool supersonic engine, this had 300 lb/s airflow, was made largely of titanium (long before any other engine) and housed accessories in a pressurized sealed box. Many aspects of the design were far in advance of normal practice. First run was on 15 December 1954, and first flight (B-47 rear fuselage pod) in 1956. Ratings then were 19,250 lb dry and 27,000 lb with full afterburner, weight being 4,120 lb. This superb engine was cancelled along with the Arrow on 20 February 1959. Orenda practically vanished, until 1966 when Orenda Ltd was formed, owned 60/40 by Hawker and United Aircraft.

Below *The Orenda 11 powered the main run of CF-100 all-weather interceptors, the Mks 4 and 5. Note the free-spinning alcohol de-icer windmill on the bullet nose.*

Bottom *Outside the Soviet Union the Orenda Iroquois was perhaps the greatest turbojet of the 1950s. This 1958 development engine, complete with afterburner, is seen en route to the testbed.*

P

Packard (USA)

Jesse Vincent built a 905 cu in V-12 of 250 hp in 1917 and used features of this in the Liberty. After the war his company naturally used the Liberty as starting point for its A-1500 and A-2500 V-12s, the designations giving the displacements in cubic inches, with typical ratings of 520 and 750 hp. In 1926 two X-2775s (basically an X-24 using 1,500 cu in cylinders with shorter stroke) were built on Navy contract for the Schneider Trophy, giving over 1,250 hp, but aircraft float trouble prevented them from reaching the start line.

In 1927, partly motivated by the Wasp, the US Navy said it would henceforth buy air-cooled engines exclusively, and Packard then fell out with the Army as well on matters of policy! The company abandoned the military market and got chief engineer Captain Lionel Woolson to develop a 9-cylinder radial diesel, the DR-980. In 1930 this was certificated at 225 hp at 1,950 rpm for a weight of 510 lb, but its smell and vibration proved insuperable drawbacks. It did, however, set an unrefuelled duration record of 84 h 32 min (Bellanca, 28 May 1931) which has never been broken. In World War 2 the company mass-produced the Merlin as the V-1650, one of which, very carefully uprated, holds another record—for piston-engine speed at 517.06 mph.

Captain Woolson and a German engineer, Prof Herman Dohner, designed the world's first practical aircraft diesel engine, the Packard DR.980, first run in 1928. The arrangement of direct air inlets on one side of the cylinder head with exhaust to the rear on the other side, was followed in the Guiberson of two years later. The DR-980 established a fuel cost of 1c/mile in many aircraft; sadly Woolson was killed (in a diesel Verville Air-Coach) through no fault of the engine.

Piaggio (ITALY)

This company obtained licences to build the Jupiter in 1928 and Gnome-Rhône radials in 1930, using these as the basis for a family of radials with 7, 9, 14 and 18 cylinders. Most had Mercury-size cylinders 146 × 165 mm, and all were named Stella. Examples were the P.VII (7-cylinder) of 400-510 hp, the P.IX, P.X and P.XVI (9-cylinder) of 625-700 hp, the P.XI and P.XIV (14-cylinder) of 700-1,100 hp, and the P.XII and P.XXII (18-cylinder) of 1,500-1,700 hp. Since 1953 the company has made Lycoming and Rolls-Royce engines under licence.

Pobjoy (UNITED KINGDOM)

David R. Pobjoy was perhaps the only proprietor in the jealousy-ridden British industry between the wars to be liked by everyone. He designed a series of beautifully engineered small radials from 1926, forming Pobjoy Airmotors & Aircraft Ltd in 1930. All his production engines had 7 cylinders and ran at high rpm, with spur gears to a propeller shaft above the crankshaft. First came the Pobjoy P, run in 1927 and type tested in 1928. Of only 25 in diameter, it had cylinders 2.385 × 3.425 in and gave 65 hp at 3,000 rpm. Next came the R, later named Cataract, with bore increased to 3.0625 in, giving 75 (later 80) hp at 3,300 rpm. In 1934 came the Niagara, with enclosed valve gear and other refinements, rated at 95 hp at 3,650 rpm (Niagara III), followed in 1937 by the Niagara V with bore further increased to 3.21875 in (192 cu in) to give 130 hp at 4,000 rpm. Pobjoy went public in 1935, Short Brothers buying a controlling interest in 1936. Pobjoy designed the world's first APUs and auxiliary gearboxes, made by Rotol, and was designing his first post-war engines when he was killed in an air crash in 1947.

Potez (FRANCE)

This prolific aircraft company formed an LEM (engine design laboratory) in 1928, subsequently

A close-cowled Pobjoy Niagara I in the first production Short Scion of June 1934. All Pobjoys were amazingly quiet.

producing a range of 3, 6 and 9-cylinder radials originally derived from Anzanis. In 1935 Henri Potez planned a range of 4, 8 and 12-cylinder inverted engines all using the same air-cooled 2-valve cylinder of 125 × 120 mm. The 4-D ran before the war, but marketing did not begin until the company was reformed in 1949, with works at Argenteuil. The 12-D never went into production but a 6-D was added, this being the first with direct injection (305 hp supercharged, the supercharger lying flat on top of the engine). Biggest of the range, the 8-D 30 was geared and rated at 500 hp at 2,650 rpm. In 1963 the company was acquired by Avco and its dwindling aero-engine business terminated.

Power Jets (UNITED KINGDOM)

In 1928 Flight Cadet Frank Whittle wrote a thesis at the lately opened RAF College in which he described how gas turbines and jet propulsion would free aircraft from the existing limitations on flight performance. In January 1930 he applied for the first patent for a turbojet, but despite his reasoned pleading neither Air Ministry nor British industry showed the slightest interest. In May 1935 he was approached by former Cadet R.D. Williams and J.C.B. Tinling, who got bankers O.T. Falk & Partners (after a favourable independent assessment by M.L. Bramson) to put up capital for Power Jets Ltd, formed in March 1936. BTH (British Thomson-Houston) were contracted to produce sets of drawings, and in June 1936 a contract at cost plus was placed for most parts of the WU turbojet. Ideally many sets of major components would have been made and tested individually, but with practically no money or resources all that could be done was build a single complete engine. Even then, while the double-sided centrifugal compressor and single-stage axial turbine were far beyond all previous experience, the requirements for the combustion chamber exceeded normal practice by a factor of more than 25. Whittle visited the British Industries Fair and was practically laughed off every stand until he found in Laidlaw, Drew & Company a firm at least prepared to tackle the colossal problem.

The first engine was ready on 12 April 1937, after total expenditure of about £3,000. On the first run Whittle opened the main fuel valve at about 2,300 rpm. At once the unit accelerated away violently out of control, the combustion chamber suddenly glowing bright red in places and the engine emitting a giant rising shriek like an air-raid siren. This

*Known as the WU, from Whittle Unit,
this famous engine was the first true aero gas
turbine in the world to run. It no longer
exists, even in a much-rebuilt form.*

happened persistently, each time causing a rapid
exodus by all save Whittle, who finally realized that
overnight pools of kerosene were collecting in the
chamber and on each test the engine ran away until
this pool had been consumed. This and hundreds of
other problems progressively reduced the U engine to
'a running heap of scrap', but it was the only Whittle
engine in existence until the first run of the greatly
improved W.1X on 14 December 1940, this time with
a little Air Ministry money. The incredibly myopic
official view changed in summer 1939 to a surprised
belief that Whittle might have the basis for a practical
engine, and a contract was placed for a flight engine,
the W.1, and the Gloster E.28/39 aircraft. The W.1
ran on 12 April 1941 and the E.28 flew on 15 May, the
engine giving 850 lb at 16,500 rpm, for an installed
weight of 623 lb.

In 1940 Gloster was given a contract for prototypes
of the F.9/40 (later Meteor) twin-jet fighter and
Power Jets was authorized to design its W.2 engine to
be made under direct Ministry contract by the Rover
Car Company. This led via the W.2 Mk 4 to the
W.2B, slightly differing examples of which were
made by BTH, Rover and Power Jets themselves,
which by this time (late 1941) was a substantial
organization with over 500 personnel. In retrospect it
would have been far better had Power Jets been
allowed to grow normally until it could take its place
as a full member of the British industry and war effort.
Sadly, once it could be seen that Whittle's engine was

actually going to work, there arose strong feelings of
jealousy and animosity whose seeds were planted at
the most crucial period of the war. In the case of Rover
it was not so much fear of competition as a progressive
deterioration in personal relationships, originally
triggered by Rover's long delays (they blamed
Whittle) and refusal to abide by their agreement not
to undertake redesign, the culmination being Rover's
unauthorized redesign of the W.2B into the W.2B/26
with straight-through instead of reverse-flow
chambers. By November 1942 the two parties were
hardly speaking. S.G. Hooker of Rolls-Royce,
appalled at the situation, got his company to replace
Rover (see RR).

On 22 July 1941, when just two Whittle engines
existed, Colonel (later General) A.J. Lyon USAAF
and D. Roy Shoults of GE visited MAP to discuss gas
turbines. The result was that on 1 October 1941 the
W.1X and a complete set of W.2B drawings were
flown to Washington, and what happened next is
recorded under GE. Power Jets themselves continued
with the W.2B/500 with longer turbine blades, typically giving 1,850 lb thrust at 16,750 rpm, followed by
the W.2/700 with a new compressor diffuser with
cascade-assisted 90° bends, and later with a more
reliable compressor rotor from GE in the first reverse
flow of gas-turbine technology. The W.2/700 was by
1944 giving 2,485 lb with airflow 47.15 lb/s from the
same size engine as the 1941 W.1. Power Jets also
carried out pioneer testing of afterburners (including

Right *This was the actual Power Jets W.1X which was shipped to the USA on 1 October 1941. It was the only bench engine Whittle had at that time! Inlet out of sight at left, reverse-flow cans fed by long curved pipes, and jetpipe attached on right.*

duct augmentation for the proposed M.52 supersonic aircraft), variable nozzles, thrust spoilers and reversers, and even a ducted fan run behind a W.2/700.

P&W Canada had a struggle to get started with the PT6 in 1958, but 14 years earlier Power Jets designed almost the same engine, to give an initial 250 hp and later twice this amount. According to former PJ designers it was Hayne Constant who 'threw it on the scrap heap', claiming nobody wanted such an engine. It was picked up by Coventry Climax Limited, which was eager to get into gas turbines, and called the C.P.35, but no real effort was put behind it. The neat turboprop had an amidships inlet to the double-sided 1st-stage compressor; the second impeller was single-sided, and the jetpipe was split on each side of the free-turbine optional rear drive.

In April 1944 the Ministry of Aircraft Production under Sir Stafford Cripps nationalized Power Jets, took over its assets and thereafter turned it into the National Gas Turbine Establishment. It was replaced by Power Jets (R & D) Limited which, forbidden to make anything, merely administered the Power Jets patents. Whittle was showered with honours and told not to rock the boat.

Pratt & Whitney (USA)

In many respects the world's biggest aero-engine company, United Technologies' Pratt & Whitney was formed in 1925 as Pratt & Whitney Aircraft Company. More than a year earlier the president of Wright Aeronautical, Frederick B. Rentschler, had begun to see serious danger signals in the bland indifference of the company board to proper engineering research and development to keep the products competitive. In fact, Wright was in a very strong position, and there was no immediate problem. But Rentschler talked with General Patrick, Admiral Moffett and Chance Vought about the prospects and decided that, despite a gloomy general picture, there would be a significant near-term market for a superior engine if one could be created. Such an engine, a 9-cylinder air-cooled radial, could indeed be created, and Wright chief engineer George Mead and chief designer Andy Willgoos were the men to do it.

Rentschler resigned from Wright on 21 September 1924. Six months later he took his battered briefcase to see the Pratt & Whitney division of the Niles-

Right *The first Pratt & Whitney Wasp, photographed at Christmas 1925.*

Bement-Pond company, of Hartford, Connecticut. As a boy Rentschler was familiar with P&W machine tools and other engineering products; now he was simply asking them for a quarter of a million dollars to build a prototype engine, plus a further full million should that prototype lead to production!

Most investors at that time would have thought the proposition laughable, but Rentschler got his backing, and Pratt & Whitney Aircraft was incorporated on 23 July 1925. Rentschler became president, Mead vice president and Willgoos chief engineer. They had to move fast to beat the Wright Simoon for promised US Navy orders. There was never any doubt about how the new engine would be arranged, though the only numerical demands were for 400 hp within a weight of 650 lb. Willgoos bent over the drawing board in his garage in Montclair, New Jersey, and by 3 August tons of tobacco had been cleared from a building on Capitol Avenue, Hartford, and the P&W executives moved in to set up an experimental machine shop. Work went on round the clock, and at last the first Wasp engine was ready on Christmas Eve. It took a whole day to get it rigged for testing; then on 29 December 1925 it was started for the first time, and within a few days was delivering 425 hp, for a weight just under 650 lb.

Its cylinders were 5.75 × 5.75 in, giving 1,344 cu in. Each barrel was machined from a steel forging with exceptionally thin close-pitch fins. The cast aluminium head was screwed and shrunk on, and contained integral rocker boxes for the enclosed single inlet and exhaust valves, with telescopic covers over the pushrods. The crankcase was assembled from identical front and rear halves joined on the centreline of the cylinders, each an aluminium forging. The two-piece crankshaft ran in two roller bearings sharing the load evenly, and was fitted with vibration dampers; it also permitted the use of a one-piece master rod. Mead devised a rotary induction system with a low-pressure blower giving good distribution from a single carburettor, and Willgoos added a fully accessible rear cover for the auxiliaries. Later the Wasp design was described as 'clean as a hound's tooth'.

The first Wasp romped through the Navy 50-h test in March 1926, the final reading being 415 hp at 1,890 rpm, and in fact never flew. The second Wasp flew on 5 May 1926 in a Wright F3W-1 Apache, in the hands of BuAer test pilot Lieutenant C.C. Champion Jr. The result was dramatic, and soon Wasps were flying in Curtiss, Boeing and Vought aircraft with startling improvements in every aspect of performance. Production engines were delivered from December 1926, output by February 1927 reaching 12 per month. Meanwhile, back in January 1926 Mead and Willgoos had completed the design of the

The first production A-series Wasp, now in the National Air and Space Museum, Washington. The author spent 683 hours behind later R-1340 Wasps, and they never missed a beat.

1,690 cu in Hornet, which the Navy wanted to replace the heavy Packard in torpedo and bomber aircraft. The first Hornet ran in June 1926 and the Navy picked the direct-drive version for the T4M torpedo-bomber whose installed powerplant weight was exactly halved in comparison with the Packard (a matter of 3,000 lb in total aircraft weight) whilst having a top speed 15 mph higher. It was at this point that the Navy announced it would buy no more water-cooled engines.

At the 1928 National Air Races Major-General Fechet, Army Air Corps chief, was so impressed by Boeing's new XF4B fighter that he ordered it on the spot (ahead of the Navy), getting his P-12A version in the summer of 1929. By this time Wasps and Hornets had set many world records, were in massive production for the Army and Navy and powered 90 per cent of American commercial transports, adding up to an amazing 60 per cent by value of the total business reported by the nation's 25 leading engine firms. So on 16 July 1929 Mrs Rentschler dug the first spadeful for a totally new plant out in East Hartford, costing $2 million. With 400,000 sq ft under one roof and 30 test cells it seemed fabulous; little did anyone think P&W would soon expand to 6 *million* sq ft!

Left *The R-1830 Twin Wasp was one of the great engines of all time. Two applications alone—19,000 B-24s and 10,000 C-47s—took a fair number, and it went into 87 other types as well.*

Above *First of the oddballs was the R-2060 of 1931, with 20 liquid-cooled cylinders. Results were so poor that the company went off liquid cooling until 1935.*

By 1930 P&W had enlarged the Hornet to 1,860 cu in and offered a whole family of Wasps and Hornets including versions with supercharging and geared drive. This had taken the Wasp up to 500 hp, and to compete in the 300/400-hp market the Wasp Junior was born, with cylinders 5.1875 in bore and stroke (985 cu in). In 1927 Leonard S. 'Luke' Hobbs had been hired as a research engineer, and one of his first jobs was to study the prospects for twin-row engines. Use of existing parts led to the R-2270 (cu in), run in May 1930 but used for research only. But in 1932 two production engines appeared, the R-1535 Twin Wasp Junior with 14 Junior cylinders, and the R-1830 Twin Wasp with 14 cylinders of a new size 5.5 in square. The 1535 began at about 600 hp and gradually developed to 825 in World War 2. Compared with a Hornet it gave similar power, was heavier and more expensive but had lower installed drag, and in single-engined aircraft offered better pilot view. It was never important, though Hughes picked a standard example for his privately funded racer which in 1935 gained the world absolute land-plane speed record. The 1830 was a different case, starting life at 750 hp and by 1936 with improved fuel giving 1,000 hp. This was a splendid engine, made possible by the company's policy of using only 2 valves per cylinder, and it opened a market unreachable with a single-row engine.

This was the bright side. Conversely, the company's still quite small engineering team had pro-

duced the R-2180 Twin Hornet, which gave 1,400 hp on test in 1935, and from 1931 had expended masses of nervous energy on the R-2060, a totally different engine with 5 banks each of 4 small liquid-cooled cylinders which developed up to 1,116 hp for the Army. Moreover, in 1928 Rentschler, Boeing and Vought had created the giant United Aircraft and Transport Corporation, adding Hamilton (propellers), Sikorsky (chiefly for Igor Sikorsky's talent), Stearman (small aircraft) and Standard Steel Propeller (to buttress Hamilton). Later the airlines had to be hived off under a new 1934 Act to leave United Aircraft; but the still enormous group took most of Rentschler's time, and his efforts to find a successor as P&W president from within the company proved troubled.

In 1936 Mead and Hobbs set the company on course again, concentrating on the R-1830 with single-stage and 2-stage or turbo-superchargers, and embarking on the big R-2800 Double Wasp with 18 cylinders 5.75 × 6 in (2,804 cu in) to replace an 18-cylinder project of 2,600 cu in. They dropped the Hornet and R-1535 and also the R-2180, even though it was ready for production. Against Hobbs' belief that they were nowhere near the limits of the engine they knew—the air-cooled radial—the pendulum then began to swing the wrong way again with pressure from the Navy, which wanted 2,300 hp, and the Army, which saw the XP-37 pursuit prototype as indicating lower drag from liquid-cooled engines. For the

Navy P&W started the air-cooled X-3130 but found they were on the limits of air cooling and substituted the liquid-cooled XH-3130 in April 1937. In the same month Mead visited England and returned full of enthusiasm for the high-speed multi-cylinder liquid-cooled sleeve-valve engine. He persuaded Douglas to fit such engines buried in the wings of a new high-speed bomber, and then contracted with the Army for the X-1800, with H-24 twin-crankshaft layout, the 1800 denoting the power (capacity was 2,240 cu in). Work begain in 1938. In November of

the same year the XH-3130 was terminated and replaced by the H-3730, an even bigger sleeve-valve H-24 engine aimed at 3,000 hp.

By 1939 Mead, a sick man, was directing the troubled liquid-cooled programmes from his home. Rentschler viewed the situation with disquiet, and Hobbs, who in nine months had brought the R-2800 to 2,000 hp and readied it for production, decided to investigate the cooling and installed drag of a 3,000-hp engine with four rows of 7 conventional radial air-cooled cylinders. The results were surprisingly

Left *One of the greatest wartime Double Wasps was the R-2800-59, used in later P-47Ds. The Dash-63 was a -59 with Scintilla instead of GE ignition. The small black unit projecting at far right, under the carburettor, is the water injection regulator.*

Below *Army pressure forced a return to liquid cooling, but the XH-3130 gave way in 1939 to the much more powerful X-1800 with 24 sleeve-valve cylinders. This is a X-1800-SA-G without aftercoolers.*

Below right *For comparison, this is an H-3730, basically an X-1800 with bigger cylinders and 2-stage superchargers and aftercoolers. It was potentially a 4,000-hp engine.*

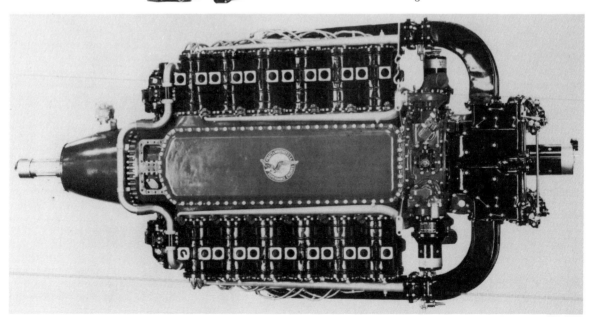

encouraging. In mid-1939 Mead resigned, and Luke Hobbs took over responsibilty for engineering. A year later, by which time the order-book had been multiplied by 10 (half of it British and French orders), the XF4U-1 prototype, powered by an early R-2800, reached 405 mph in level flight, a world record for a military aircraft. Soon afterwards General 'Hap' Arnold visited Hartford. Rentschler told him bluntly the liquid-cooled engines would be too late for the war, but that P&W could build an equally powerful conventional engine that could make it. Arnold reputedly said, 'Now we're getting somewhere'. The R-4360 Wasp Major team was formed the next day, and the liquid-cooled engines were cancelled.

Back in 1939 business had actually been depressed but then the French orders had paid for a plant addition of 280,000 sq ft. Then 'The British Wing', paid in sterling, added 425,000 sq ft from June 1940. US funds added another 375,000 sq ft, and this was just the start. In late 1940 Ford, the first licensee, began a process which duplicated not only the R-2800 but also the complete Hartford plant. Buick built the R-1830 and the R-2000 Twin Wasp (for the DC-4/C-54, with 5.75 in bore); Nash-Kelvinator, the R-2800; Chevrolet, the R-1830 and later the great R-2800 C-series fighter engine with wet ratings up to 2,800 hp; and Continental and Jacobs built, under license, R-1340s and R-985s, respectively.

Meanwhile, Hartford grew to 3 million sq ft, another 2 million was added in satellites in Connecticut, and finally in 1943 a new plant was built outside Kansas City, Missouri, even bigger than Hartford

and put to making the new C-series Double Wasp, a colossal challenge because the workforce had to be trained from scratch and much of this largely new engine had never been mass-produced anywhere and involved new techniques. The mighty R-4360 had already passed its first tests by June 1942, and later in the year was qualified at the unprecedented rating of 3,000 hp. An unusual feature of this engine, apart from its 'corn-cob' layout, was that it had a one-piece crankshaft and split master rods. By VJ-Day it had been qualified at 3,500 hp.

On that day P&W's giant order-book was slashed close to zero. Thousands took a vacation, the first for four years, while in the boardroom Rentschler, 'Jack' Horner and general manager Bill Gwinn added up the horsepower of the wartime engines and found it came to 603,814,723 from 363,619 engines. No other company has ever come anywhere near this. But on Gwinn's office wall there soon hung a cartoon: Gwinn, aboard a battered sailboat, asks 'See anything, Jack?' Skipper Horner, looking to the horizon, replies 'Nope'. In fact, the traditional engines were to go on for a long time, with even the R-2180 being designed as a new replacement for the Twin Wasp, with 14 Wasp Major type cylinders. Called Twin Wasp E, it was certificated as a bolt-on power egg (a first for P&W) for the DC-4; but by this time nobody wanted DC-4s and the only application was 18 Saab Scandias. The last throw was the R-4360 compound, which went beyond 4,000 hp. The vital application was the DC-7, and Wright won because the Turbo-Compound was lighter and simpler.

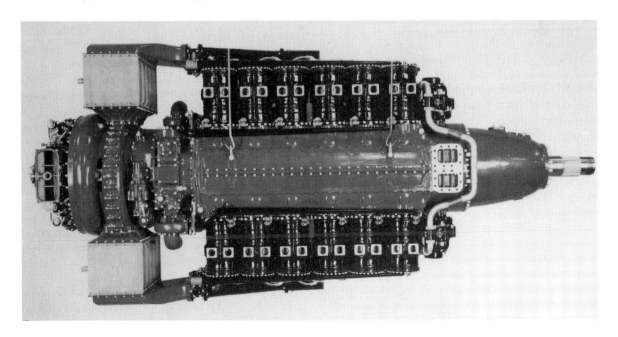

Gas turbines

Hobbs had studied turboprops in 1939, and from 1940 collaborated with MIT on the PT1 in which the propeller was driven by a turbine fed by eight 2-stroke free-piston diesel gas generators. Effectively the giant firm had nothing in this new field, and it set 1950 as the time by which it must learn all about it and then leap-frog past its rivals. It learned a little immediately after the war building 130 Westinghouse J30 turbojets, but the big project that tided P&W over the hard pre-1950 period was the decision of the US Navy to buy the British Nene turbojet.

The Navy stipulated the first J42 must roll off the Hartford line in November 1948, in time for the F9F programme. It also had to be Americanized, be engineered for mass-production (it was not being

Left To Luke Hobbs' relief the liquid-cooled engines were cancelled and replaced by the R-4360 Wasp Major. This was still no small challenge, but one that could be mastered before the end of the war. All 28 cylinders got proper cooling.

Below left Britain ignored the potential in the Nene and Tay, but the Hartford engineers took these engines to 8,300 lb with afterburner in this J48-P-5A. Airflow was 130 lb/s.

Bottom left A production T34-P-3, the 5,700 shp (6,000 ehp) turboprop which powered the C-133A. Weighing 2,670 lb, it was the company's first gas turbine.

Below Mockup of the XT45 two-spool turboprop, the faltering seed from which stemmed all today's big engines from Hartford.

made in quantity in England) and be able to come from a US source. Gwinn headed a delegation to Derby in August 1947, the first J42 (a Nene with US accessory systems) was running in March 1948, and it was in production with a wet rating of 5,750 lb by October. Hobbs had insisted on the need for future development, as a result of which the two firms jointly produced the Tay and J48, the latter following the J42 with ratings up to 6,250 lb, or 8,750 lb with afterburner.

In June 1945 Hobbs had begun development of a large single-shaft turboprop, the PT2, funded by the Navy as the T34. Made largely of stainless steel, it had a 13-stage compressor handling 65 lb/s at pr of 6.7, a cannular combustor and 3-stage turbine. It entered service in modest numbers in 1958 at ratings of 6,000 and later 7,500 ehp. Once design of this engine was complete, in March 1946, studies were begun for a successor. Hobbs was determined to start the leap-frog process, and the choice fell on an axial turbojet of 8,200 lb thrust, the JT3. By May 1947 the JT3-6 was in detail design, with a compressor of 6 pr with a constant diameter of 36 in. The Navy rejected it—ironically, because Westinghouse offered the J40 sooner and cheaper. In January 1947 a study for a related 10,000-hp turboprop showed that with the desired high pr of 8 the cranking power needed to start the engine would be exorbitant, and flow through the compressor would be very poor. This led in July 1947 to the idea of splitting the compression between two spools, each turned by its own turbine via concentric shafts (an idea investigated by P&W's R.G. Smith

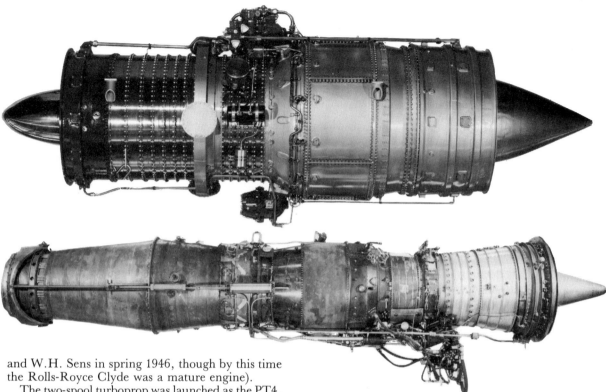

and W.H. Sens in spring 1946, though by this time the Rolls-Royce Clyde was a mature engine).

The two-spool turboprop was launched as the PT4, funded by the Air Force as the XT45. Components were made in late 1947, while parallel studies were made of the two-spool JT3-8 turbojet, which by March 1948 was supercharged by adding two stages at the front, giving 10,000 lb thrust. During 1948 Air Force interest hardened on a high-compression turbojet for long-range bombers, and the XT45 was terminated in September 1948. Instead the J57-P-1 specification was written, and the shops began building two actual engines, the X-176 (JT3-8) and X-184 (JT3-10). Even as they were being built, rig testing indicated poor performance, mainly because of the very small HP compressor blades, poor turbine disc design and excessive weight.

Mechanical design came under Willgoos (it was his last engine, for he died after shovelling snow in March 1949) and aerodynamic design under Perry W. Pratt. In February 1949 they worked out a scheme to redesign the JT3 in a wasp-waisted form, the rotor discs having a constant diameter but the casings tapering towards the HP end amidships, to give higher efficiency, better sealing and cut 600 lb off the weight, besides improving the arrangement of accessories to give a more compact installation in fighters or in bomber pods. Pratt got the go-ahead to redesign in May 1949, but it was decided to complete both the barrel engines, the X-176 running on 28 June 1949

and the X-184 in February 1950. Both confirmed the poor performance. The wasp-waisted design threw up its own problems, notably with bearings and compressor-blade vibration, but with massive effort these were 'trampled to death'. The first redesigned JT3 ran on 21 January 1950, flew under a B-50 in March 1951, completed a 150-h test in J57-3 production configuration in November 1951, powered the 8-jet YB-52 on 15 April 1952, and with afterburner took the YF-100A beyond Mach 1 on its first flight on 25 May 1953.

This was probably the most important engine in the world since 1945. It initially gave 10,000 lb dry or 15,000 lb with afterburner yet, because of its pr of 12.5, it set totally new standards in jet fuel economy. Almost certainly its use in short-range fighters was mistaken: they would have done better with a less-economical engine of half the J57's weight of about 5,000 lb with afterburner. For the B-52 and many other long-range aircraft the J57 opened up possibilities previously only dreamed of, not least being the design of the 707 and DC-8 commercial jets using civil JT3C engines. The latter entered scheduled service on 26 October 1958 with TBO of 800 h, and were developed later to a dry thrust of 13,000 lb for a weight of 3,495 lb, almost 1,000 lb lighter than the original B-52 engine, and with TBO of 14,120 h. The JT3/J57 was made of steel (later versions had a titanium LP compressor), with 9 LP and 7 HP stages, 8 flame tubes in a cannular chamber and 1 + 1 turbine stages; airflow was 164-187 lb/s, pr 11 to 13.8, and highest rating 13,750 lb wet (P-43s and P-59s in B-52F/G and KC-135As) or 19,600 lb with afterburner (P-420 in F-8J). In 1951-60 P&W made 15,024, and Ford in Chicago a further 6,202.

The JT3 fulfilled every hope in overcoming P&W's late start in gas turbines, and fortuitously the failure of the rival J40 gave it the US Navy market which had not been expected. Obviously it was not an end but a

beginning, and from a host of possibilities the first derived engines were the T57 and JT4. The former still remains the most powerful turboprop built outside the Soviet Union, rated in the 15,000-hp class. It added extra LP turbine stages driving a front reduction gear to a giant HamStan B48 single-rotation propeller with four hollow-steel blades; its application, the C-132 military airlifter, was cancelled in 1957. The JT4, funded by the military as the J75, was a JT3 scaled up to 249-256 lb/s airflow, retaining similar pr despite having only 8 LP stages. Bulk and weight were increased by only some 10 per cent, but typical ratings were 17,500 lb dry and 26,500 lb with afterburner. Though heavier and less fuel-efficient it far outsold the Rolls-Royce Conway

The 15,000-hp PT5 (XT57) turboprop compared with a J57 bomber engine. The XT57 was flown in the nose of a C-124C Globemaster II.

One of the two JT9A-20 (J91-P-1) engines tested in summer 1957. The high design Mach number is reflected in the nozzle diameter of 65.6 in, compared with 35.0 in maximum for the engine. Airframe-mounted nozzle ejector flaps were bigger still.

in 707s and DC-8s, and as recently as 1984 former afterburning fighter J75s were being rebuilt to power TR-1s.

In 1951 work began on an indirect-cycle nuclear powerplant based on a PWR (pressurized-water reactor) driving a ducted fan of 13.1 ft diameter and 67.6 ft long. In 1953 this was replaced by a powerplant based on a single molten-salt circulating-fuel reactor providing heat for six J91 turbojets each of almost 6 ft inlet diameter. When the WS-125A nuclear-powered bomber was cancelled in 1957 the indirect-cycle powerplant went with it, though P&W then turned to a new scheme with four J58s modified to higher pr and fed from two or even four reactors, to fly in a 500,000-lb Convair testbed in 1965. This aircraft, the NX-2, was finally abandoned in 1961.

The J91 was a major programme for the USAF aimed at the CPB (chemical-fuel bomber) WS-110A and then the NPB WS-125A. It had to be a big single-shaft turbojet stressed for March 3, fed from a variable inlet and with a giant convergent/divergent nozzle with ejector flaps of 72.5 in diameter. The chemical-fuel engines were a series of JT9s, 9Bs and 9Cs (J91-P-1 variants), all with a 9-stage compressor of pr 7, annular chamber with 8 ring cans each with 6 burners, and a 2-stage turbine. Only two experimental engines, X-287 and 291, were built (in spring 1957). They were run at 24,500 lb dry and 35,000 lb with afterburner. The JT9-5A series had dual-fuel afterburners burning either JP-5 or HEF-3 boron fuel, ratings being 28,700 lb dry, 41,500 lb with afterburner and 44,000 lb with water injection as well. In the event, GE won the CPB contract, but the JT9 provided a valuable basis for the later (but mainly unrelated) JT9D commercial engine and, especially,

the J58 as noted later. The nuclear J91, the JTN9, was never built, though a quarter-section of the heat-transfer radiator was tested.

The X-287, JT9A-20, had proved on test to be aerodynamically brilliant. Its first two compressor stages, with long transonic blades, were then scaled down to 53-in diameter, handling 450 lb/s airflow, and substituted for the first 3 LP stages of the JT3 turbojet. Almost the only other modifications were to add a third LP turbine stage and a new fan duct which in the B-52H was bifurcated at the rear to discharge supersonic air from left/right nozzles, having a cross-section resembling a banana. The result was the JT3D turbofan, funded by the USAF as the TF33, which almost overnight in 1960 not only fended off all competition from the Conway (which had a bypass ratio far lower than the JT3D's 1.4) but did it with a simple modification to the existing turbojet which could be carried out by the operator using a kit sent from Hartford. The conversion had only marginal effect on weight but increased thrust by over 35 per cent, reduced fuel burn by 15-22 per cent and cut 10 dB from take-off noise. (Despite this, today operators of JT3D airliners are faced with costly modifications to meet severe new FAA Stage-2 noise rules.) Production of this pioneer turbofan ended in 1985 at about 8,600, most of them not conversions. Late (P-7 and PW-100) models have 7 LP stages and rating of 21,000 lb thrust.

The next spin-off was a scaled-down turbojet, the JT8/J52, funded by the US Navy in 1954 and produced for missiles and Navy aircraft at 7,500-11,200 lb thrust. Compressor and turbine stages were 5 + 7/1 + 1, and though the planned commercial JT8 never found a market, the J52 will remain in production for the A-6 at least through 1986. In sharpest contrast the derived JT8D turbofan, produced in 1960-61 to power the 727, has had hardly any military sales (apart from the Volvo Flygmotor RM8 (*qv*), yet it has probably been the most profitable gas turbine in history, with 12,000 having logged 245 million hours at the time of writing. Compared with the JT8 the LP spool was changed to have 2 fan and 6 LP stages, driven by a 3-stage turbine, airflow being roughly doubled from 136-143 lb/s to 315-331 lb/s, pr being around 16. From 1970 the 9 cans and burners were modified to reduce smoke, and subsequently a refanned JT8D-200 series was marketed (for the highly successful MD-80 series) which takes thrust from the 14,000-17,400 lb bracket up to 20,000-21,700 lb, with reduced sfc and noise.

In the story of Garrett on an earlier page, reference is made to Randolph Rae and his attractive proposal to use liquid hydrogen to fuel highly supersonic, ultra-high-flying aircraft. On 15 June 1955 a parallel contract was awarded to United Aircraft's research division. Before the year was out the USAF was well into its gigantic Suntan project; Colonel Appold, who headed it (and later ran the C-5A on behalf of the Air Force) reckons the cost was not less than $250 million, but it was so disguised and super-secret there is no way of getting an 'accurate' figure. On 17 February 1956 P&W chief engineer, Perry W. Pratt, collected current hydrogen knowledge and began talking with Lockheed's 'Kelly' Johnson about a U-2 successor. William Sens, who accompanied Pratt, outlined an engine cycle in which the gas was burned downstream of the turbine, figures being: thrust 4,500 lb at 100,000 ft; diameter of nacelle 61 in; and sfc 0.75. One of the young J75 engineers, Richard J. Coar (today president of Pratt & Whitney), was pulled out

to organize a hydrogen deal with the Air Force; it took just one day, and one sheet of paper. Suntan multiplied, the next development being a J57 rebuilt to run on the intensely cold liquid; it was dramatically shorter than before. By August 1956, with Coar as project engineer, the world's first 'clean sheet of paper' liquid hydrogen engine was being built. It could not receive a legitimate designation, so P&W looked at the Engine Order No (703040) and extracted the digits 304. The diagram (p. 120) shows how the fuel, at 18K (minus 255°C), was pumped at high pressure through a heat exchanger containing over five miles of stainless tube with 2,240 furnace-brazed joints. Here a heat-exchange rate of 21 MW (21,000 kW) heated the hydrogen from 18K to just over 1,000K, at which temperature it expanded through a remarkable turbine with 18 small stages putting out 12,000 hp. This drove through a reduction gear to the multi-stage fan. Downstream of the turbine some hydrogen was burned, to give 1,500K; the rest was burned downstream of the turbine. Total nacelle weight was 6,000 lb, and thrust at 100,000 ft was 4,800 lb, with sfc of 0.8. The 304 began testing on 11 September 1957, running successively on nitrogen, gaseous hydrogen and liquid hydrogen. Later engines had 5 instead of 4 compressor stages, with sea-level thrust of 13,500 lb. But Suntan, like its name, faded. It encountered opposition from many quarters, and even Kelly Johnson became convinced this fuel was a non-starter (though since he retired Lockheed-California has published hydrogen-powered TriStars and similar studies). The lasting benefit was a vast national capability in the technology and supply of liquid hydrogen, which benefited the space programme.

Johnson himself re-embraced petroleum, and in

The prototype JT3D-1 turbofan of December 1958, which was just preceded by the USAF-funded TF33 version. This can claim to have been the first big turbofan, and it transformed the 707 and DC-8.

Above *The JT8D has been described as 'the world's most profitable engine' (a cynic might say because the spares cost so much). By year 2000 JT8Ds may have logged over 800 million hours, probably an all-time record for any engine.*

Left *Rated at 21,700 lb, the JT8D-219 is the most powerful of the refanned JT8D engines used in the MD-80 airliners.*

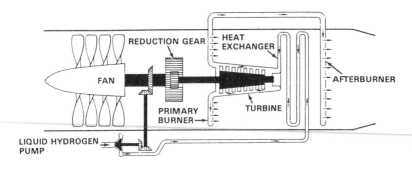

Left *The Hartford engineers were probably better fitted to tackle the highly secret Project Suntan engine than any other team aywhere, but the liquid-hydrogen pump and heat exchanger were new challenges. The 23,500-rpm pump had oil lubrication! The heat exchanger, of 71.6 in diameter, transferred heat at a rate that would 'heat 700 six-room houses'.*

Above *At Mach 3.2 almost all the thrust pulling the SR-71 'Blackbird' along comes from the inlet system, a little being added by the white-hot nozzle. In between comes this J58, which just gets in the way (but you need it for take-off).*

Right *The Pratt & Whitney '304' looked almost normal, but a fitter dismantling it would soon be amazed. The 6 variable stators show this to be the first 304-2 (June 1958) with 5-stage fan geared from the 18-stage hydrogen turbine. Two were to power the 164 ft 10 in Lockheed CL-400 at 99,500 ft.*

1958 P&W accepted the challenge of providing propulsion for his A-12/SR-71 family at up to Mach 3.35. The resulting JT11B turbojet, funded as the J58—an out-of-sequence Navy even number— is a single-spool engine made of refractory materials, with much of the supersonic compressor delivery bypassed through six large pipes; special JP-7 low-volatility fuel is used, with chemical-reaction ignition of the afterburner, and lube oil must be preheated before each engine start. HamStan provided the advanced control system for the engine, afterburner, inlet, nozzle and complex arrangements of secondary airflow doors.

In 1957 what is now P&W Canada began design of the JT12 (J60) small turbojet with 9-stage compressor (c50 lb/s, pr 6.4) rated in the 3,000-lb class. This was taken over by the parent firm, about 1,900 being delivered. From this the JFTD12 (T73) helicopter engine was derived by adding a 2-stage free turbine giving an output of 4,500 or 4,800 shp.

The JFTD12A-5A (T73-P-700) was a 4,800 shp helicopter engine produced by adding a 2-stage free power turbine downstream of a JT12 (J60) turbojet.

Chronologically the next step forward was selection by the US Defense Department of the GD/Grumman F-111 version powered by the JTF10A. If one includes work on derived engines by SNECMA this was the first afterburning turbofan, with 3 fan stages rotating with 6 LP, 7 HP, 8 flame tubes, and 1 + 3 turbines. The initial production version, the TF30-P-3, had airflow of 233 lb/s at 17.1 pr, thrust being 10,750 lb or 18,500 with full afterburner. Severe problems were experienced in matching the engine and aircraft, the inlet system needing redesign. Later the F-111 engine was developed into the P-100 version rated at 25,100 lb, yet at 4,022 lb marginally lighter than the P-3. Unaugmented versions rated at 11,350-13,400 lb powered early A-7s, the final model (P-408) in this series transferring the third fan stage to the LP compressor. In the F-14A the P-412, with a different afterburner nozzle, suffered prolonged trouble requiring repeated redesign and various 'fixes' such as armour to contain burst compressors; after 12 years the P-414A appears to have solved the main problems, but this damaged P&W's reputation with the Navy and future F-14s are GE-powered.

The company also suffered its share of problems with the next-generation fighter engine, the JTF22, designed and developed as the F100 by the Government Products division at West Palm Beach, one of the four divisions of the 1976-restructured Pratt & Whitney Group of UTC (United Technologies Corporation), which previously had been the Florida R&D Center. With this engine steel suddenly took a back seat, the dominant materials being titanium, high-nickel alloys and even more exotic materials. There are 3 fan stages, a 10-stage HP spool with 3 variable stators at the upstream end, an annular combustor and 2 + 2 turbines with directionally solidified blades. Airflow is 228 lb/s at 24.8 pr and dry and augmented ratings are 14,670 and 23,830 lb respectively. To some degree the problems stemmed from the arduous operating conditions and the USAF wish for very long service life with reduced ownership costs. A fairer picture of F100 performance is the fact that current engines fly 1,800 mission cycles without even hot-section refurbishment.

The F100 is the basis for three derived engines: the PW1115 (basically an unaugmented F100) of 15,000 lb thrust, the PW1120 (a turbojet of 20,600 lb thrust with the original fan replaced by an LP compressor driven by a single-stage turbine) and the PW1130 or F100EMD which is a growth version rated at 27,410 lb. Of these only the PW1120 has a market, initially in the Israeli Lavi and in rebuilt F-4 Phantoms.

In 1961-63 P&W fought and lost the propulsion contract for the aircraft that became the C-5A. Undaunted, it was selected by Boeing to power the 747, the resulting JT9D being the first of today's giant commercial turbofans. First run was in December 1966, first flight (B-52E) in June 1968, and maiden flight of the 747 followed on 9 February 1969. The JT9D is made of titanium, high-nickel alloys, stainless steel and other advanced metals, and in the latest models the HP turbine blades are single-crystal. The original JT9D 3 which entered service on 21 January 1970 had a 1-stage fan (1,510 lb/s), 3 LP, 11

Pratt & Whitney

Right *This TF30-P-6, for an A-7A, shows the distinctive ribbed casing that was also a feature of the related SNECMA TF306.*

Right *The F100 almost had a corner of the market with the F-15 and F-16, but caused headaches almost as prolonged as those of Navy TF30s. Today it is probably the best and most reliable fighter engine in service in the world.*

Right *Today's era of giant wide-body aircraft was launched by the JT9D engine, which has now been developed into this 7R4 model with a new 97-in fan. Spinning at only 3,530 rpm this gulps air at 1,695 lb/s on take-off.*

Pratt & Whitney wanted to make the PW2037 demonstrably the most fuel-efficient engine in the world, and made some rash promises of financial compensation to Delta Airlines if it failed to beat the competition by 'between 7 and 8 per cent'. This is impossible, but P&W vice-president Tad Domegala says he is optimistic they won't have to pay out!

HP, annular combustor and 2 + 4 turbines, weight being 8,608 lb and thrust 43,500 lb. Since then more than 2,800 JT9Ds have flown about 75 million hours with ratings up to 56,000 lb, the latest models having 4 LP compressor stages and weighing about 9,100 lb. On paper the PW4000, announced in December 1982, reads like an advanced JT9D, but in fact this is a near-total redesign at company expense to reduce the number of parts by more than 50 per cent, improve economy and promise extended life at lower cost. The first of this series, probably the PW4056 (ie, 56,000 lb thrust) is expected to be certificated in mid-1986.

In 1972 P&W began work on the JT10D to fill the gap between the JT8D and JT9D with a modern engine. A demonstrator ran in 1974 at 23,000 lb, and in 1977, following long collaboration, MTU and Fiat were brought in as risk-sharing partners. Since then the engine design has been largely started again, in the 37,400-lb bracket as the PW2037, pulling out all the stops to compete against the Rolls 535. The PW2037 has a single fan stage (78.5 in, 1,340 lb/s) rotating with 4 LP, 12 HP (all with controlled-diffusion aerofoils), an annular combustor and 2 + 5 turbine stages with active clearance control. P&W has

tried hard to market this excellent engine as the most fuel-efficient in the world, but the rival 535E4 is proving hard to beat, and sales of the 2037 were causing concern in early 1985 (a proposal to fit four in each active B-52 has not been accepted). P&W is, however, a major partner in IAE (*qv*).

In 1981 the group announced the PW3000 family of new shaft-drive engines in the 4,000-8,000 hp bracket, for use in aeroplanes and helicopters. Which of several rivals will be picked had not been announced in early 1985, but P&W hope to replace such engines as the T55 and T56 as well as power new aircraft. Pratt & Whitney is also trying not to be left behind by GE in the challenging new field of propfans. Their studies centre on engines which use a gearbox to link the small multi-stage turbine and giant counter-rotating fans. P&W's Florida plant has also been a long-term supplier of the RL10 liquid-hydrogen space rocket engine.

Pratt & Whitney Canada (CANADA)

In 1927 the RCAF was so excited by the P&W Wasp that it suggested to a group of Montreal businessmen that it should be built in Canada. James Young responded, and Canadian Pratt & Whitney Aircraft was organized in 1928. Most of its effort was devoted to service support, but in 1951 sudden demand for new R-1340 Wasps resulted in the parent company at East Hartford deciding to build a major new plant at Longeuil, Montreal, to handle all future piston production and spares. Longeuil prospered, but studies showed the business would decline. In talks with the parent United Aircraft Corporation it was decided to make Montreal the P&W company handling turbine engines for general aviation. On 1 January 1957 the first six 'new hires' for this endeavour joined the firm and formed the nucleus of a new team.

Though small turboprops were obviously the most likely replacement for the Wasp, there was the prospect of an immediate jet market in Canadair's proposed new military trainer, the Tutor. This was being designed to use the J83 or J85. The new team began designing the JT12 turbojet in the 3,000-lb class, but prospects looked interesting and with the Canadian company not yet big enough to handle it, the design and marketing team moved to Hartford, where the JT12 went ahead (see Pratt & Whitney). The team then returned to Longeuil, cut their teeth by creating a new accessory gearbox for the RR Tyne-powered CL-44, and at last, in winter 1958-59, work began on a shaft engine in the 500-hp class. Studies showed it should have a gas generator with three axial stages preceding a centrifugal and a reverse-flow overall layout, with a free power-turbine

Top *Engine of the Shorts 360, the PT6A-65A of 1,409 ehp is one of the longer and more powerful PT6 variants.*

Above *Flat-rated at 750 shp the P&WC PT6A-135A is typical of the lower-powered versions of this best-selling turbo-prop.*

Right *The P&W Canada JT15D-5 is the latest and biggest version of this popular turbofan, and it can be seen to have at last got rid of part-span shrouds (snubbers), unlike the big Hartford engines!*

shaft running away from, rather than through, the gas generator. Designated PT6, the gas generator ran in November 1959 and the complete engine in February 1960. It was way off the mark in performance and weight, but a 'Mark II' engine in July 1960 showed promise.

Features of the PT6 include: a compressor with 3 (-65 series, 4) axial stages plus one centrifugal, an annular reverse-flow combustor, an HP compressor-turbine, and an LP turbine (1-stage in early models, 2-stage in high-power versions) driving the propeller gearbox. Airflow varies from 5.9 to 9.5 lb/s, with pr from 6.5 to 10, and there are many other variables to match no fewer than 29 distinct current production models to a host of applications, with powers from 475 to 1,327 shp. Not including the T74/PT6 helicopter turboshaft and the various T400/PT6T Twin-Pac twinned helicopter engines, which have sold as well as the fixed-wing versions, some 18,500 PT6 turboprops had flown nearly 70 million hours at the time of writing, in over 144 countries!

In 1966 design began on the JT15D turbofan, which contrives to combine simplicity with advanced technology. The original JT15D-1 has a fan aerodynamically scaled to 75 lb/s from the far larger JT9D fan, followed by a titanium centrifugal compressor, annular reverse-flow combustor and 1 + 2 turbine stages, rating being 2,200 lb and weight 514 lb. The D-4 version adds an axial boost compressor stage and gives 2,375 lb, while the D-5 is a growth version with greater airflow and pr, rated at 2,900 lb. Newest P&WC engine is the PW100 (originally PT7A) turboprop, launched in 1979 and first certificated in 1983 as the PW115 (1,500 shp). Later versions are the PW120 (2,000 shp) and PW124 (2,400 shp), the latter driving a 6-blade propeller. All are 3-shaft engines with LP and HP centrifugal compressors each driven by a single-stage turbine, a folded annular combustor and a 2-stage power turbine driving a central shaft to the front gearbox. The PW115 has airflow of 14.3 lb/s and pr 10.9, other versions exceeding these values, typical weight being 841 lb.

PZL (POLAND)

PZL is the association of Polish aero and engine producers, but before the war it had different divisions one of which licensed such engines as the Pegasus and Walter Minor. In 1952 responsibility for manufacture and support of Soviet piston engines was transferred to what is now PZL Kalisz, which still makes the Ivchyenko AI-14 and Shvetsov ASh-62 (as the ASz-62IR). PZL-Rzeszow makes the Glushenkov engines as the PZL-10 series and Isotov GTD-350. The PZL-3 is a Polish 7-cylinder radial derived via the LIT-3 from the Ivchyenko AI-26; without change in capacity or rpm it has been uprated to 600 hp and is available in direct or geared forms. In the 1950s Wiktor Narkiewicz designed the WN-3 radial with 7 smaller cylinders (135 × 134 mm, 13.4 litres), rated at 340 hp at 2,500 rpm and made in small series; the WN-4 was a helicopter version. Today PZL also licenses former Franklin opposed air-cooled engines (*qv*).

R

Ranger (USA)

When the Fairchild Airplane Manufacturing Corporation was incorporated in 1925 Sherman M. Fairchild had already launched the design of an in-

Picked for several new transports, including the BAe ATP, the PW100 is a 3-shaft engine of advanced design. One unusual feature is the use of 2 centrifugal compressors running at different speeds.

Top *A typical Ranger, the 6-390-D was rated at 150 hp. Virtually all Rangers had accessories high at the back and the camshaft drive at the front.*

Above *One of the main wartime Rangers was the 6-440C-5 (military designation L-440-7), rated at 200 hp on 87-grade fuel. In 1947 the author helped sledgehammer 115 new ones, along with the Cornells in which they were installed; only the packing cases were spared under Lend-Lease rules. Location: Southern Rhodesia.*

line air-cooled engine. This was designed in 1926, built in 1927 and marketed in 1928 as the 6-370, this giving the number of cylinders and capacity. This engine had steel barrels, shrunk/screwed aluminium cast heads with al-bronze seats for the 2 valves, an underhead camshaft driven from the front (with a magnesium cover serving as the sump) and a built-up aluminium crankcase with 7 main bearings. A distinctive feature was the mixture supply manifold

which had a T-junction leading to 12 further T-junctions each serving a single cylinder on one side and 2 on the other. This basic configuration lasted through all the successor engines, all of which had slightly bigger cylinders. The sixes went through 390 and 410 to a final 441 cu in (4.125 × 5.5 in), the 6-440 being rated at 175 to 200 hp at 2,450 rpm, depending on sub-type and fuel grade, for a weight of about 376 lb.

The name Ranger was introduced in 1930, Ranger Engines Division becoming a subsidiary of Fairchild Engine & Airplane Corporation on the latter's formation in 1936. By this time a market was appearing for the V-770, an inverted V-12 with 773 cu in from cylinders 4 × 5.125 in and with fork/blade conrods. Fair numbers were produced in 1941-45 at 520 hp at 3,150 rpm, geared and supercharged, weight being 730 lb. Post-war business dwindled, ending with 2-cylinder airborne generating sets for large aircraft. The J44 is dealt with under Fairchild (*qv*).

Rateau (FRANCE)

Auguste Rateau had a factory at La Courneuve, on the road from Paris to Le Bourget. Famed for his steam turbines, and from 1917 for his turbocharger, he also studied gas turbines and during World War 2 he designed the SRA.1 axial bypass jet, run in 1947. It had a 4-stage LP and 12-stage HP compressor, the latter being bypassed by much of the flow, all on one shaft. Thrust was claimed to be 1,200 kg. After 1950 the all-steel SRA.101 Savoie followed, with 10-stage compressor and again with 12 pipes bypassing much of the flow round the 12 combustors. Planned for 4,000 kg, it was again unsuccessful. This may have been due to fundamental errors in aerodynamic design which surfaced when, in 1960-64, the company tried to sustain a gigantic lawsuit against other turbojet companies and operators (it lost).

Reaction Motors (USA)

Reaction Motors Incorporated was formed in 1941 by four members of the American Rocket Society, and soon found itself involved in major development programmes for engines for test vehicles, missiles and aircraft. Of the latter the first was the 6000C4 (6,000 lb thrust, 4 chambers). Developed from 1943 for the XS-1 (later X-1), it used gas pressure to feed ethyl alcohol and lox to 4 stainless-steel thrust chambers each with its own electric ignition. Chambers could be fired separately or in any combination. As the XLR11 it was built in small numbers for X-1 variants and the D-558-II, later engines having turbopump feed. In 1955 RMI was awarded the propulsion contract for the X-15 hypersonic research aircraft. While the LR99 engine was being developed, the first

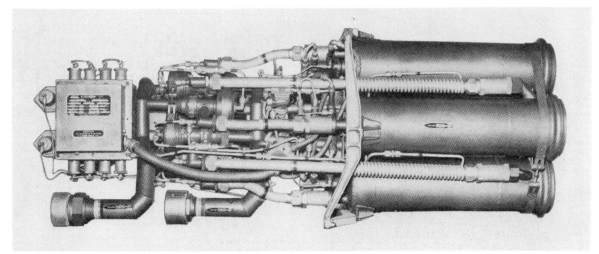

two X-15As were flown with twin LR11-RM-5s each rated at 8,000 lb at SL. The first LR99 Pioneer engine was delivered in May 1960, and on 8 June exploded during a ground run in the No 3 aircraft. As a result the first flight with the LR99 was made by No 2 aircraft on 15 November 1960. No larger than the LR11, the LR99 had a diameter of 40 in, length of 72 in, weight of 910 lb, and burned anhydrous ammonia and lox at 10,000 lb/min to give 50,000 lb thrust at SL and 57,000 lb at high altitude. It was fully throttleable and could be repeatedly shut down and restarted in space. RMI also developed small helicopter tip-drive rockets. In 1958 the company was taken over by Thiokol Chemical (now Morton Thiokol).

Regnier (FRANCE)
Starting as licensees of the Gipsy Major and Six, this

small company quickly introduced successive modifications until the engines were fresh designs. The only one of importance was the 4L, with cylinders 120 × 140 mm (6.3 litres), first run in 1936 at 90 hp and developed post-war at 147 hp at 2,340 rpm, or with 7.2 compression as the 4L.02, giving 170 hp at 2,500 rpm. Unusual features included steel cylinders with baked varnish protection, Y-alloy heads held by long bolts into the crankcase, and special metal/plastics sealing rings for the 2 cylinder joints. Many sub-types existed, differing mainly in the arrangement of accessories. Regnier was absorbed by SNECMA in 1947.

Renault (FRANCE)
This great car company was started by the brothers Fernand, Louis (the designer) and Marcel Renault,

Above *RMI (Reaction Motors Inc) deserve to be better remembered as the company that powered the first supersonic aircraft: Bell X-1 series, Douglas Skyrocket and (initial tests) NAA X-15. The four chambers put out jets at 2,760°C at 6,182 ft/s, to give specific impulse of 192. The engine was 56 in long and weighed 210 lb.*

Left *The Regnier 4L was made in large numbers by SNECMA, though at over 150 kg it was heavy for its power.*

and they completed their first aero engine in January 1908. They chose a 90° V-8 with air-cooled cylinders 70 × 120 mm, producing 35 hp at 1,400 rpm. By January 1909 they had increased bore to 90 mm, giving 6.1 litres, the new rating being 55 hp at 1,600 rpm. Also run in 1909 were a 25-hp V-4 and a 45-hp water-cooled upright 4-in-line. The latter had twin overhead valves, flywheel and (unlike the other engines) direct drive from the crankshaft, and proved a dead end. All the other engines had cast-iron cylinders of F-configuration, with a lateral projection on the inner side providing for an overhead exhaust valve worked by pushrod and rocker and an inlet valve on the underside, driven by a direct pushrod. Petrol feed was from a carburettor mounted low down on one side of the engine. The camshaft was strengthened to drive the propeller, though it is doubtful if a big 4-blade propeller at 650 cruising rpm was any better than a 2-blader driven off a 1,300-rpm crankshaft. Tens of thousands of Renaults were delivered by 1918, many of them directly derived V-12s and with various cylinder sizes up to 125 × 150. Some were pushers, and these had a crankshaft-driven cooling fan and tight-fitting air casing. They were heavy, very conservatively rated, had to be

Above left *This V-4 of 25 hp of 1909 was the first Renault production aero engine. The propeller was driven off the (half-speed) camshaft. This engine has a high carburettor.*

Top *Last of the Renault 4-in lines, the 4P was made in small numbers even after the company was absorbed into SNECMA in 1946. Like the 6Q it was known as the Bengali (with suffix 4 or 6 depending on number of cylinders), a typical rating being 140 hp.*

Above *During World War 2 Renault made many hundreds of 6Q (Bengali 6) engines of 220 or 240 hp, but an RAF raid on Billancourt in 1943 brought output almost to a halt. Note the twin carbs, and magnetos above the crankcase.*

decoked every 20 hours and fully overhauled every 50-70 hours, and, being partly fuel-cooled, were amazingly inefficient; but their reliability made up for all the rest.

The chief post-war engines were the 300-hp 12E and 12Fe and 420-hp 12K, all with 125 × 150 cylinders. The company persisted throughout the 1920s with big V-12s, both air and water-cooled, but in 1928 wisely went for the low-power market with the 4P upright 4-in-line. A crude but effective 90-hp unit, it was developed into the 4Pbi and Pci inverted

Left *Variously known as the Renault 12S and SNECMA 12S, this 600-hp inverted V-12 was developed from the German Argus AS 411, last and most powerful of the Argus range.*

Right *The start of Rolls-Royce in aviation was the Admiralty order for 'the 200-hp engine'. This Eagle VIII of 1917 gave almost double that output, establishing a tradition of ceaseless development.*

Below right *The original wartime Condor seen here was basically an enlarged Eagle with a horizontal sump and 4-valve cylinders. In June 1921 Rowledge came from Napier and turned the Condor into a much more modern engine.*

engines; bore was then enlarged from 115 to 120 mm, stroke remaining 140, to give the 4Pdi Bengali of 6.3 litres (same as later Regniers), starting life at 145 hp at 2,350 rpm. There followed a series of inverted 4-and 6-in-lines spurred by Coupe Deutsch racing, which by 1940 had also led to remarkable light fighters. Most of the latter had the 450-hp 12RoI inverted V-12, but the final Caudron prototype, the C.R.770, had the prototype of a new inverted V-16, the Renault 626 of 800 hp. Raymond Delmotte took off on the first flight as the German troops approached the test airfield at Guyancourt; the long crankshaft had insufficient bearings and broke after a few minutes, and Delmotte had to turn back. The 500-hp R468 inverted V-12 ran at 500 hp in April 1940, but its C.R.780 airframe was never completed.

During the war Renault produced 4P and 6Q in-lines as well as the Argus As 411. In 1945 the latter was further developed, culminating in the Renault (SNECMA) 12T of 625 hp at 3,300 rpm. Renault directors were taken by the Paris Communists in 1945 and never seen again, the company being nationalized into SNECMA.

REP (FRANCE)

Pioneer aviator Robert Esnault-Pelterie powered his 1907 and 1908 monoplanes by ingenious engines often described as 7-cylinder radials. In fact they were in effect 2-row fan engines, the front row having 4-cylinders and the rear row 3, all mounted round the upper part of the crankcase. Cylinders were 85 × 90 mm (30-hp) or 85 × 95 (35 hp). Each cast-iron cylinder was bolted to the aluminium crankcase, pistons

being very thin-walled steel with the gudgeon pin held by a cast pivot screwed into the piston head. The cylinders each had one valve, set alternately in one of three positions: inlet (pushed down), both ports closed, and exhaust (pulled up to open the surrounding collar). In 1909 REP even sold several 60-hp 14-cylinder 4-row engines.

Rolls-Royce (UNITED KINGDOM)

One of the world's best-known companies was formed in 1906 by Henry Royce, who struggled from humble beginnings to become an engineer, and the Hon Charles S. Rolls, a wealthy sportsman who took an honours engineering degree at Cambridge. Their first product was the Silver Ghost car, in production in 1907, but the actual running of the firm was left to chairman Ernest Claremont and managing director (commercial) Claude Johnson. Royce had extraordinary foresight for the day in realizing that there would soon be a market for aero engines, but his interest turned to antipathy when Rolls was killed in 1910 when his Wright suffered inflight structural failure. In 1911 Royce became seriously ill, and for the rest of his life convalesced, from 1913 in the south of France, then at St Margaret's Bay, Kent, and finally at West Wittering, Sussex. He nevertheless remained the focal point for every technical detail of the company's products to his death in 1933, and the extra effort and travelling this caused may be imagined.

On the outbreak of World War 1 Commodore Sueter of the Admiralty accepted the suggestion of

Wilfred Briggs, head of the Air Engine Section, that the famous Derby firm should be asked to design a 200-hp aero engine, mainly as an insurance to back up Sunbeam. Briggs' assistant Lieutenant W.O. Bentley (later car designer) agreed that a good starting point would be the cylinder of the Mercedes racing car that had won the 1914 French Grand Prix, which happened to be on show in London. He did, however, later suggest the pistons should be aluminium. On the first Sunday of the war, 9 August 1914, Briggs and Bentley towed the racer to Derby. There its engine was immediately dissected by Ernest Hives, head of the experimental shop and chief test driver. Hives, Hs in company shorthand, was a man of towering character and capability. Even in 1914 with his small hand-picked team he exuded confidence to solve problems quickly.

Hives alone would have put Rolls-Royce in a special position, and he was the perfect foil to Royce's humourless insistence on painstaking development to increase power and reliability by the methods which today are taken for granted but which were widely regarded with bemused disbelief in 1914. But there

was an extra ingredient present from the start, injected by Johnson. This was a belief in RR superiority and indeed infallibility. Like the Admiralty the War Office wanted the firm's resources applied to improving Britain's pathetic showing in aero engines, but did so by asking it to make Renault or RAF engines under licence. Johnson's reply was typically uncompromising: 'Such a plan would yield nothing but mountains of scrap. . . I would go to prison rather than agree to it'. It is rather surprising that he did agree to using features borrowed from Mercedes, and in fact both licensed Renault and RAF.1A engines were made.

Down on the Kent coast Royce had fewer inhibitions. As a true engineer, his objective was the best product. He spent nearly a month evaluating alternative engine layouts, and took particular care not to infringe any Mercedes patents. On 8 September he finally settled on a water-cooled V-12 with cylinders 4.5 × 6.5 in, 1,238 cu in, to give the required 200 hp at 1,600 rpm, with 0.64 geared drive. Each cylinder was a steel forging, held by studs through the flanged base on the cast aluminium crankcase. Inlet and exhaust ports were forgings welded on opposite sides of the head, the two overhead valves being driven by single part-exposed camshafts. At the rear were bevel drives to the centrifugal water pump, main oil pump, two 12-cylinder magnetos serving two plugs per cylinder, a tachometer and hand-cranking dog clutch. Two Claudel Hobson updraught carburettors were used (the handbooks called these 'Rolls-Royce') each serving one cylinder bank. The V-angle was 60°, and each crankpin was driven by one master and one articulated rod. Water jackets were welded steel pressings, each needing its own piping. The only unusual feature was the reduction gear, of the epicyclic type.

Royce strove for refinement and reduced weight, having wooden models of many parts sent from Derby and returned with detailed instructions for modifications. The prototype '200-hp engine' was started in March 1915, and Hives soon had it running at 1,800 rpm and giving 225 hp. Meanwhile, Royce approved two further engines, a 6-in-line with 4 × 6 in cylinders, and a V-12 with cylinders 4 × 5.75 in (867 cu in). Prototypes appeared in winter 1915-16, the V-12 being designed at Derby by R.W. Harvey-Bailey. Johnson then named all three engines respectively the Eagle, the Hawk and the Falcon.

Almost all the effort went into the Eagle, which Royce, increasingly delegating work to assistant A.G. Elliott, kept improving. His stream of letters, memos and drawings to Derby, outlining design philosophy, test procedures and many other matters, were considered so important—and so far ahead of

This photograph of the Eagle XVI was taken from the drive end, showing the casing of the epicyclic reduction gear. Cast aluminium throughout, it was a very clean engine.

other companies—that in December 1915 the directors had them printed in a limited edition of 100 and bound, one copy being shown only to selected young engineers to ensure the tradition of excellence. A basic philosophy, though expressed in very different words, was 'flog it until it breaks, think about the failure and modify the part'. By this means the Eagle power rose from 225 hp to 266, 284, 322 and by March 1917 to 350 hp in the Eagle VIII, all continuous ratings at 1,800 rpm. Higher powers were available for 5 minutes at 2,000 rpm. Except for the Eagle I and IV all had 4 carburettors, the Mark VIII having two at each end, and later engines had four 6-cylinder magnetos. A typical 350/375-hp Eagle VIII weighed 847 lb.

The Eagle, the 75/105-hp Hawk and 205/285-hp Falcon all had important wartime applications, and in the immediate post-war era the Eagle was the leading British engine in large military and civil aircraft, making the first direct crossing of the Atlantic and the first flights to Australia and South Africa. Later Eagles were cleared for an overhaul life of 100 to 180 hours, then exceptional figures. In the 1920s it faded against competition from the Jupiter and Lion, but firmly established RR's reputation in aero engines.

In 1917 design was started on an Eagle scaled up to 5.5 × 7.5 in cylinders (2,138 cu in) to power the V/1500 bomber. This was the Condor, and though it was not ready in time for its original application its power of 550/600 hp found plenty of other uses. The big cylinders each had 4 valves, though still with a single camshaft on each bank, and the twin carburettors were mounted low down on each side at the midpoint of the engine. Another new feature was an electric starter. In the early 1920s A.J. Rowledge, from Napier, was given the job of redesigning this obsolescent engine, and the resulting Condor III of 650 hp had fork-and-blade conrods, spur reduction gear, and improved valve gear and accessories. But Trenchard wished all RAF engines to be capable of dispersed manufacture by car firms; RR still insisted they alone could make their engines, and when he read this Trenchard scrawled across the page 'No more Condors'.

Royce had already thought about a new engine, so that when in 1925 Lieutenant-Colonel L.F.R. Fell of the Air Ministry asked British firms to beat the Fairey Felix, the groundwork had been done. It is recorded that Royce drew preliminary outlines with his stick on the sand at West Wittering, leaving Rowledge to do the design in detail. The new engine was called the Eagle XVI, though it had nothing in common with the Eagle but water cooling, for it was an X-16 with 4

Rowledge also played a big part in designing the Kestrel, with cylinders in cast blocks. This is an early production example, geared but unsupercharged (and with Hucks starter dogs on the front of the prop shaft).

banks of small 2-valve cylinders supercharged by a double-sided gear-driven blower with 4 volute outlets, one to each bank. This was run at some 500 hp in mid-1926, and a bigger Eagle XX was planned, but both were then dropped in favour of a modernized V-12 to compete more directly with the Felix.

Elliott helped at Wittering, while at Derby Jimmy Ellor, top supercharger expert at Farnborough, joined the firm to help Rowledge design this new component. The big difference in the F, as the new engine was called, was that each bank of 6 cylinders was a monobloc aluminium casting, with thin steel open liners of 5 × 5.5 in size (1,296 cu in) resembling those of the Puma rather than the Hispano or D-12. The first type test, at 490 hp at 2,350 rpm, was gained by the compact F.10 direct-drive engine in spring 1927. In 1928 various geared Fs appeared, and in 1930 it was named Kestrel. By this time the company had begun to develop its own aircraft installations at Tollerton, later moving to Hucknall airfield, north of Nottingham, and this growing establishment played a major role in almost all subsequent engines. One of its first tasks was to produce a ram air inlet, first used on the R (see below), which added some 10 mph to Kestrel aircraft. During the 1930s full supercharge, automatic boost control, 87-octane fuel and salt-cooled exhaust valves all resulted in improved Kestrels, the final marks being rated at 745 hp at 14,500 ft at 3,000 rpm. Kestrel deliveries were 4,750, just topping the wartime total of Eagles (4,674).

In the late 1920s some work was done on diesels, aided by Harry Ricardo and the RAF, and two compression-ignition Condors were type-tested in 1932 and flown. The future, however, lay with the monobloc petrol engine, and to power large flying boats an enlarged F was designed in June 1927 as the H, later named Buzzard. This geared and supercharged engine had cylinders 6 × 6.6 in (2,239 cu in), and initially weighed 1,460 lb, rating at sea level being 925 hp. Few were built, but in 1928 it formed an admirable base for the R engine to contest the 1929 Schneider Trophy. Basil Johnson, brother of the late Claude, was opposed to any form of racing (and in fact resisted increased expenditure on aviation by what he considered a car company). Air Ministry pressure secured from Royce an assurance that an engine would be forthcoming. Royce schemed a narrow V-16, but there was not enough time and the only way to meet the schedule was to use the Buzzard as starting point. Highly stressed parts were strengthened, the crankcase, valve gear and accessories redesigned to improve streamlining, and a giant double-sided supercharger fitted with a ram induction pipe between the blocks. The first endurance run was in May 1929, the promised 1,800 hp being held for 1 h at 2,900 rpm. F.R. Banks then advised on special fuel mixes, and the 1929 race was won using 78 per cent benzol and 22 per cent Romanian naphthenic petrol, plus 2.5 cc TEL per gallon. Power was 1,900 hp for a weight of 1,530 lb. On the night before the race there was a frantic cylinder-block change on Flying Officer Waghorn's S.6, the fitters being rounded up from

The Buzzard was essentially an enlarged Kestrel; only 100 were built, but they powered a great variety of aircraft. Its chief importance was in providing the basis for an engine to contest the Schneider Trophy.

Southampton's pubs. Waghorn was not told until he won the race; his son still has the failed piston.

After Lady Houston had donated £100,000 to enable the RAF to contest the 1931 race, the government having declined to do so, Rolls realized they were faced with the biggest challenge yet posed by a petrol engine. The go-ahead was delayed until January 1931, so all that could be done was to work around the clock on the R. Rpm were increased, supercharger gear ratio raised and the air inlet enlarged; fork/blade rods were replaced by articulated rods, main bearings gave prolonged trouble, and all main moving parts were again strengthened. Valve springs lasted minutes only, and at one time oil consumption was 112 gal/h, but in August 1931, using a further Banks (Associated Ethyl) special brew and pure castor oil lubricant, 2,350 hp was held at 3,200 rpm at 67.2 in manifold pressure, weight being 1,630 lb. Spare blocks were not needed, and the trophy was won for keeps on 13 September 1931. Soon afterwards Banks came up with 30 per cent benzol, 60 per cent methanol and 10 per cent acetone, plus 4.2 cc/gal TEL, and with this 2,783 hp was obtained at 72.3 in at 3,400 rpm. Nothing remotely like this had been done before, and when the company said 1931 equalled 'five years of normal development' it was probably an understatement.

By 1927 the company was thus firmly in the aero engine business, and had begun to engage engineers such as Lovesey, Rubbra and Dorey. They studied competitive weaknesses, and the most obvious one seemed to be the massive weight and other drawbacks of the water cooling system. In 1928 a special F was designed to use steam cooling. As the heat dissipator (radiator) in such a system is full mainly of steam instead of water it weighs much less, and condensing the steam gets rid of about 30 times as much heat as cooling the same flow of water by the typical 17°C. The steam-cooled engine, which became the Goshawk, ran very well at around 660 hp, but flight testing revealed various problems, such as the impossibility of pumping condensate at near boiling point back from a condenser lower than the header tank, as it inevitably had to be in low-wing monoplanes. So in 1935 Rolls-Royce followed the US lead and went for ethylene glycol cooling. A 97/3 per cent mix with water boils at sea level not at 100°C but at 164°, so heat can be dissipated through a very hot but smaller and lighter radiator.

Thus only 24 Goshawks were made, though they were flown in 12 different aircraft. Another smaller engine was the Peregrine, a 1939 modernized Kestrel with a downdraught carburettor and many new features. In the Whirlwind it was rated at 885 hp at 3,000 rpm at 15,000 ft, on 87-grade fuel. Almost the

Above *The R was by a wide margin the 'No 1' racing engine of its era, and Campbell and Eyston later used them to set land and water speed records. Note the enormous ram inlet duct and supercharger, which by October 1931 was generating a manifold pressure of 72.3 in—yet an R could idle at 475 rpm.*

Right *Big downdraught carburettors on an unpainted Kestrel-size engine show this to be a Peregrine, used in the F.9/37 and Whirlwind twin-engine fighters. It introduced progressive boost control, each increment of throttle-lever travel giving a corresponding variation in power.*

same cylinder blocks were used in the Vulture, described later.

Where next, then, after the 1931 Schneider win and world speed record? The Air Ministry declined to fund a new engine, and in December 1932 the RR board voted to pay for one. Three years later the Hawker board did the same in tooling up to make 1,000 Hurricanes, powered by the engine that RR were to develop, and in neither case could any of the people who took those courageous decisions have had any idea how important they would prove to be in 1940. In early 1933 Royce, then weak and frail, decided to go ahead with a conventional V-12 with cylinders 5.4 × 6 in (1,649 cu in), to begin life at 750 hp and develop to 1,000. Called the P.V.12 (private venture), it almost went ahead with the inverted layout, which gave better pilot view, but there were lubrication problems and Royce disliked anything new that was not essential. He died on 22 April 1933, on the very day the last P.V.12 drawing was completed.

This Merlin II typifies the initial production version of what some would call the world's most famous engine. Take-off power was 890 hp, and full-throttle power 950/990 hp at 12,250 ft. A similar engine was boosted in 1937 to 2,160 hp, or a remarkable 0.621 lb/hp.

Elliott had deserted Royce sometime earlier in order that the design team could be concentrated in Derby where Rowledge and Colonel Barrington had been the leaders. The first P.V.12 ran on 15 October 1933.

Seldom has an engine proved so disappointing for so long. The double-helical reduction gears had to be replaced by plain spur gears, and the ambitious aluminium monobloc casting that included the upper crankcase and both cylinder blocks kept cracking through the water jackets. By 1941 such cracks were being repaired by the hundred, but in 1933 it meant paying for a new casting and losing weeks. Somehow a P.V.12 was got through a type test in July 1934 at 790 hp at 2,500 rpm, weight being 1,177 lb, and another flew in a Hart on 12 April 1935. By this time Air Ministry support was forthcoming, and the engine was named Merlin. The Merlin B introduced ramp-type (semi-penthouse) detachable heads with the twin inlet valves at 45°, but continued cracking forced the use of separate cylinder blocks, in the Merlin C. This failed a type test at 955 hp in May 1935, but did complete a 50-h civil test in December and the Merlin F was committed for production with minor changes as the Merlin I, for the Fairey Battle and Hawker Hurricane.

Within days it was realized the ramp head potential was not being achieved, and it was at once replaced by the Merlin G with an improved flat head cast integral with the block, with all 4 valves parallel. About 180 Merlin Is had been delivered before the G came on the line in early 1937 as the Merlin II, type-tested at 1,030 hp at the full rpm of 3,000. It required complete redesign of the noses and engine controls of the Hurricane and Battle. Hives, by this time 'King' of the Derby works, sanctioned a racing Merlin in 1937 which ran 15 h at 1,800 hp and for short bursts at 2,160 hp at 3,200 rpm at 27-lb boost, on fuel of about 100 octane. This amazing achievement confirmed that the 27-litre Merlin would probably be adequate to beat the much bigger German engines. The Mk X with two-speed blower went into production for the Whitley and Halifax, and in 1938 'Doc' Hooker joined the firm and quickly discovered how to make a major improvement to the performance of the supercharger.

The Battle of Britain, however, was won by the original Mks II and III, but operating on the very limited quantities of 100-octane fuel first brought to Britain in bulk in June 1939. A boffin at the RAE, Miss Shilling, hit on the brilliantly simple idea of adding a metal diaphragm with a small calibrated hole inside the float chamber of the carburettor to keep the engine running under sudden negative-g in combat, and thus stay on the tail of a direct-injection 109 even if it went into a dive. Hooker's new supercharger resulted in the Mk XX (Hurricane), 45

(Spitfire) and many related engines for bombers. The conversion of the Manchester into the Lancaster was made possible because of the prior existence of a Merlin 'power egg' developed for the Beaufighter II, all ready to bolt on. Production was more than quadrupled by new factories at Glasgow and Crewe, and by Ford Motors at Manchester. A plan involving Ford at Dearborn fell through, to be replaced by Packard Motor Corporation, where the first V-1650, a totally redrawn Americanized Merlin, was run in August 1941. Half a dozen development V-1650s were made by Continental.

In mid-1940 a demand for a high-altitude Merlin for the Wellington VI was expected to be met by turbocharging. Instead Hooker developed a 2-stage supercharging system, with an intercooler, the first blower being derived from that of the Vulture. The result was the Merlin 60-series, which doubled power at high altitudes, adding 10,000 ft to the Spitfire's ceiling and 70 mph to its speed. General engineering improvements were introduced continuously and piecemeal, so as not to disturb production, so that by the war's end the Merlin had been transformed. In 1945 special Merlins were giving up to 2,780 hp (over 100 hp per litre) at 36 lb boost, running on fuel with monomethyl aniline added. Merlin production figures were Derby, 32,377; Crewe, 26,065; Glasgow 23,647; Ford 30,428; Packard 55,523.

Ever alert to new technology, Rolls-Royce began research with sleeve-valve cylinders not later than 1933, and in 1934 built a complete sleeve-valve Kestrel. This ran satisfactorily, though it was used to provide design experience and never flew. With the knowledge gained, Rowledge spent 1935 designing the aptly named Exe. This totally new engine was a single crankshaft 90° X-24 with tiny cylinders 4.225 × 4 in (1,348 cu in) with sleeve valves and pressure air cooling. First run in September 1936, it was initially rated at 920 hp at 11,000 ft but was soon giving 1,200 hp with plenty more to come. One of the smoothest and most trouble-free engines imaginable, it was intended for Royal Navy aircraft, starting with the Barracuda, because of its air cooling. It was always a low-priority oddball and in the vast expansion of 1938 was cancelled, but it was so popular with the Hucknall test pilots they used the Exe-Battle as their 'hack' communications aircraft until at least 1943. Exe-type design never stopped, and the Pennine was planned as a much more powerful (2,500 hp) successor for post-war civil aviation, only to be overtaken by the gas turbine.

More sustained effort was applied to the Vulture. This was the direct route to higher power, in having 4 Peregrine blocks (2,592 cu in) arranged in the same X-24 form with 60°/120° spacing and one crankshaft. Design began in September 1935, testing

started in May 1937 and the Vulture II was type-tested at 1,800 hp in August 1939. Production began in January 1940, but endemic conrod failures forced maximum rpm to be reduced from 3,200 to 3,000 and finally to 2,850. Despite this, the take-off rating was raised to 2,010 hp at 9 lb boost on 100-grade fuel in March 1941, but in-service accessibility was poor, inflight fires and other snags made the Vulture unpopular, and Hives decided his engineers could

Below *The air-cooled sleeve-valve Exe was Rowledge's last design, and it never gave a moment's trouble. The company Exe-Battle flew several hundred hours as a regular liaison aircraft after the engine itself had been cancelled!*

Bottom *In stark contrast, Vultures gave so much trouble that 97 Squadron (Manchesters) were invariably grounded, and known in Bomber Command as 'the 97th Foot' (for foreign readers, 'Foot' meant an infantry regiment).*

spend their time more profitably on better engines. At termination in April 1942 only 508 production Vultures had been delivered, all for Manchesters.

On 1 January 1939 Harry Cantrill, previously with Armstrong Siddeley, was given the job of developing the Griffon. This was to be a conventional V-12 scaled up from the Merlin to give over 1,500 hp at low altitudes in naval torpedo bombers (oddly, the obvious application, the Barracuda, was left with the small Merlin when the Exe was cancelled). Sensibly it was decided to make the new engine as compact as possible so that it might replace the Merlin in some applications. Cantrill moved the drives to the camshafts and magnetos to the front, not only saving length but almost eliminating torsional vibration in the drives. Another feature, previously tried only on RR experimental engines, was to feed oil end-to-end down the hollow crankshaft to the main and big-end bearings. The Griffon had cylinders of Buzzard/R size (6 × 6.6 in, 2,239 cu in), and started off with a lot of Merlin experience built in; yet for most of the war Lovesey managed to keep the Merlin ahead, especially at altitude. The Griffon never equalled the Merlin's specific output, but was nevertheless a fine engine, which transformed the Spitfire and Seafire (not least in rotating in the opposite direction, which reversed take-off procedures, though this attribute disappeared with contraprops, which were common on the Griffon). Later marks gave around 2,000 hp to at least 20,000 ft, maximum power being about 2,540 hp, though in such racers as the *Red Baron* Mustang Griffons have delivered 1,000 hp more than this.

Least known of the company's piston engines, because everyone was too busy to write it down at the time, the Crecy was the most advanced 2-stroke aero engine ever built. Sir Harry Ricardo began the work in about 1937, testing single-sleeve cylinders with a clever system of charge stratification from an injector in the head. One feature was the high exhaust energy, making an exhaust turbine desirable. Ricardo's results justified Derby building a complete engine, and by 1941 the Crecy was on test. A liquid-cooled V-12, it had cylinders 5.1 × 6.5 in (1,593 cu in), a little smaller than the Merlin in displacement, and with the blocks spaced at 90°. Project engineer Harry Wood recalled from a distance of 25 years that without a turbine the engine gave 2,000 hp at 2,600 rpm, and was expected to reach well over 3,000 hp with a gas turbine geared to the crankshaft. The Crecy team made extravagant claims for it, but it suffered from severe piston overheating and Hooker concluded it could never compete with the 4-strokes in output per unit piston area.

In most ways it was a more advanced engine than the last Rolls piston engine, again named Eagle. Started in December 1942, this was a 'clean sheet of paper' engine, the premises being that for reasons of flame travel bore could not exceed 6 in and that it was undesirable to have more than 12 cylinders on one crankshaft, while experience with the Exe and Crecy confirmed at least a potential advantage of sleeve valves. The result was a flat-H with 24 cylinders 5.4 × 5.125 in (2,808 cu in), the crankcase being split into left/right halves each carrying a monobloc casting interchangeable left/right, with upper and lower rows of 6 cylinders held by 28 nickel-steel through-bolts. The engineering looked good, though the 3,500 hp rating established at 3,500 rpm for the first small production series demanded 28-lb boost.

Gas turbines

Prompted by Hooker, Hives visited Whittle at Lutterworth in August 1940 and immediately broke a bottleneck in Power Jets operations by making blades, casings and many other parts in the Derby experimental shops. This got Derby into turbine hardware, though the chief scientist, A.A. Griffith, who had joined from Farnborough in June 1939, had brought with him a scheme for a turbojet as complex as Whittle's was simple. Griffith had tremendous knowledge of gas turbines, and proposed turbofans and turboprops both using his 14-stage CR (contra-rotating) gas generator which also had contra-flow of air and gas. The research unit, designated CR.1, never cured its air and gas leakages which prevented realization of the efficiency predicted.

The need to get into gas turbines and jet propulsion remained, and by 1941 Hives was seeking a link with Power Jets. After many discussions he collected a small design staff at Derby to build the WR.1 turbojet (actually as a subcontract from Power Jets), using Whittle parts except for the conservatively designed 2.6 pr compressor. The basic reason was to explore the possibility of making a turbojet totally reliable, but by the time two WR.1s had been run in late 1942 they were overtaken by events.

The scandalous situation at Barnoldswick (see Power Jets) would probably have continued, and kept British jets out of the war, had not Hooker kept Hives informed. Unlike most of British industry, who regarded Whittle as an upstart and potential competitor, RR got on well with Whittle and understood the colossal importance of his invention. So one day in November 1942 Hives, Hooker and S.B. Wilkes, chairman of Rover, met for a cheap wartime dinner at 'The Swan and Royal' in Clitheroe. After the meal, in a few short words, a deal was struck that was to change history. Hives asked 'Why are you playing around with this jet engine? It's not in your line of business'. Wilkes replied that he would like to be shot of the whole business, whereupon Hives said, 'You give us this jet job, and I'll give you our tank engine

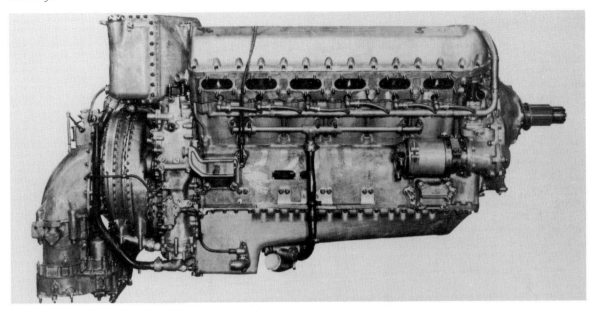

Above *Though 36 per cent larger in capacity, many Griffons were actually shorter and lower than other typical 2-stage Merlins. This Griffon 65 (Spitfire XIV, once flown by the author) developed almost 200 hp per cylinder.*

Right *Like the Exe, the Crecy liquid-cooled sleeve-valve 2-stroke seemed to be less troubled than some mature production engines. The 6-plunger injection pumps can be seen at the front of each block. The Crecy, like the 4-stroke Pennine, never flew.*

Right *Last Derby piston engine, the mighty Eagle was very much like a scaled-up Sabre, but avoided the latter's problems. This is a contra-rotating Mk 22, rated at 3,500 hp and with 3,020 hp available at 15,250 ft.*

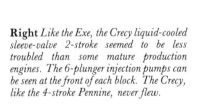

Below right *The B.37 Derwent I introduced 'straight through' chambers which transformed the appearance. This engine was made at Newcastle-under-Lyme for the Meteor III.*

factory at Nottingham'. Thus from 1 January 1943 Hooker became chief engineer at a disused cotton mill at remote Barnoldswick, which had been planned for W.2B production but which was henceforth used merely for development. To say the British jet-engine scene changed almost overnight is to put it mildly: W.2B running time was 24 h in December 1942 and it was over 400 h in January 1943. A situation that disgraced the nation was suddenly rectified, and Whittle's engine was—for the very first time, incidentally—at last in full development, with proper resources.

Among the Rover engineers taken over at Barnoldswick were Adrian Lombard and John Herriot, the latter chief test engineer and the key to making the W.2B/23 work, and 'Lom' destined later to be director of engineering until his sudden death in 1967. Rolls left off the 'W.2', without offending Whittle, and as the Rolls-Royce B.23 the engine flew in the F.9/40 Meteor at 1,400 lb on 12 June 1943. In October it was cleared to 1,600 lb and a batch of 100 delivered as the Welland I from Barnoldswick to power the Meteor I. The weight was 850 lb, and in August 1944 squadron aircraft were fitted with nozzle inserts which reduced area and increased thrust to 1,700 lb to catch flying bombs.

It had long been known that greater airflow could be handled by a redesigned Welland, and to produce the B.37 engine Geoffrey Wilde reprofiled the diffuser, the turbine being modified to accept the new

airflow of 38.6 lb/s compared with 32.4. Straight-through (B.26) combustion chambers were used, the outstanding Lucas team under Dr E.A. Watson continuing to mastermind combustion problems. The fuel system was redesigned with a barometric control governing the stroke of a swash-plate multi-plunger pump. Many new features were introduced, including expansion joints in hot parts and a recirculating oil system with oil cooler, which were to become standard gas-turbine practice. The new B.37 engine named Derwent I, ran at 1,800 lb in August 1943 and was type-tested at 2,000 lb, the design figure, in November. By this time Herriot had gone to run a new factory at Newcastle-under-Lyme where 500 Derwent Is were turned out in the second half of 1944 for the Meteor III. They raised low-level speed from 415 to 470 mph, roughly 100 mph faster than any other Allied fighter at this height.

Because of its better aerodynamic design the Me 262A was slightly faster, but there was a startling contrast between the engines. The best TBO for any Jumo 004B was 30 h, whereas the Welland and Derwent both passed type-tests of 500 h and had a service TBO of 150 h. At last combustion ceased to be a wholly dark art, and with the Derwent the fuel system, burners and chambers had reached close to the form used on thousands of subsequent engines. Turbine blades also ceased to be a limiting factor, and though the Derwent I ran at 843°C compared with 754° for the Welland (because of removal of troublesome inlet guide vanes) the Nimonic blades took it in their stride. When improved vanes were restored, the gas temperature returned to the original figure, without altering the 2,000-lb rating and reducing sfc from 1.178 to 1.083. Meanwhile Whittle's team produced the magnificent Type 16 compressor casing which kept the airflow under perfect control from the impeller tip to the entry to the combustor liner. It was applied to the Derwent II, raising thrust to 2,200 lb. The Mk III engine incorporated powerful suction pipes for boundary-layer control on the A.W.52. The Mk IV had an impeller enlarged from 20.68 to 21.7 in diameter, giving 2,450 lb, but suddenly a way was found to leapfrog all such development.

In early 1944 the Ministry requested RR to design an engine of 4,200 lb thrust, and Hooker and Lombard schemed the RB.40 (initials for Rolls-Royce Barnoldswick were used henceforth). In mid-1944 Hooker visited the USA and discovered that GE had two types of engine of 4,000 lb already running. He therefore decided to go for 5,000 lb, and under Lombard the B.40 was turned into the B.41 Nene. Whilst actually reducing engine diameter slightly, to 49.5 in, the impeller was enlarged to 28.8 in, to handle 80 lb/s, and other features were incorporated

from Derwent testing. The prototype was built in six weeks, and arranged so that inlet vanes could be added if desired. Harry Pearson failed in his efforts to get them put in for the first run, on 27 October 1944. Various snags delayed things until near midnight. Then, with virtually the entire day and night shifts watching, the engine was started. It failed to light; positioning the igniter was then a hit-or-miss affair. After several attempts Denis Drew unscrewed the igniter and, as the big Nene cranked up to speed, he lit it with an acetylene welding torch (thus was born the torch igniter). Gradually the throttle was opened, and the cheer as the needle pased 4,000 lb could be heard all over Barnoldswick. Hooker heard the Nene running as he arrived early next morning, and was told the inlet vanes had been installed in the small hours. He looked at the instruments: the Nene was now giving 5,000 lb at the same temperatures as the 4,000 lb on the previous night!

Thus in five months the Barnoldswick team, small by today's standards, had created the most powerful engine in the world. They had done it by using a superior compressor with efficiency at pr 4 raised from 74 to 79 per cent, housed in a Type 16 casing with 9 efficient outlets to 9 large Lucas chambers of

Top *With the Nene, Hooker and Lombard multiplied available thrust by a factor of 2.5 within a few months, and for very modest increases in engine size and weight (yet it was Britain's foreign competitors who made best use of the Nene and its development the Tay). This is a Nene 10, made in Montreal for the T-33AN.*

Above *Though a lash-up, the Trent was the first turboprop to fly. Even earlier, in May 1944, a Welland had been tested as a shaft-drive engine.*

the latest type. The compressor and turbine were now on separate shafts, linked by a quick-detach coupling and running in pressure-lubricated ball and roller bearings and with a small impeller added to cool the turbine bearing and disc. It was supplied by a pair of

the previous standard fuel pumps, arranged so that it could run (at reduced thrust) on either, and the pilot's throttle was rigged so that, instead of getting all the power in the last half-inch of travel, equal movements of the lever gave equal increments of thrust. Materials were Nimonic 75 for the cans, Nimonic 80 for the turbine blades, and Jessops G.18B for the disc. Weight was about 1,600 lb.

Roy Chadwick wanted to put four Nenes in a Lancaster, which would have been almost immune to interception or flak, but by this time the rot had set in at the political level and the Nene was left to the Soviet Union, Hispano-Suiza, Pratt & Whitney, RR Canada, Commonwealth Aircraft and, later, the Chinese. Whittle came to see its first tests, and at the 'Swan and Royal' that evening everyone bewailed the lack of any application. Someone (thought to have been Whittle) suggested scaling it down to fit the nacelle of the Meteor. Herriot or Lombard did the calculation on the tablecloth and got the amazing answer of 3,650 lb thrust! As they were toiling to increase Derwent thrust from 2,200 to 2,450 lb this seemed too good to be true. Hooker did a very quick sum and said 'We've got a 600-mph Meteor'. Hives had just got the production line turning out Derwent Is, but he did not specifically forbid Hooker's suggestion, so on 1 January 1945 the drawings began for the Derwent V, a 0.855-scale of the Nene. On 7 June the first engine started a 100-h test at 2,600 lb, and it soon reached 3,500 lb, weight being 1,250 lb. Meteors set world speed records at 606 mph in November 1945 and 616 mph in September 1946, the latter with 4,200-lb engines with Nimonic 90 blades.

In late 1943 Hooker concurred with Griffith's arguments and decided to work on turboprops for aircraft speeds of some 400 mph. By early 1944 an interim scheme had been drawn. Called the RB.50 Trent, it was effectively a Derwent II with a flexible quill shaft (to isolate the flimsy engine from feared vibration) to a reduction gear and 95 in Rotol 5-blade propeller. These parts were designed by A.A. Rubbra, then chief designer at Derby, and Lionel Haworth, who was already working on the RCA.3 3-spool axial jet of 80 lb/s airflow (never built). The Trent ran in June 1944, and Eric Greenwood made the world's first turboprop flight at Church Broughton with the Trent-Meteor (converted Mk I *EE227*) on 20 September 1945. On landing, Greenwood throttled back, the propellers going into the special flat zero-pitch used to facilitate starting; the Meteor dropped like a stone, and only quick full power saved a crash. This aircraft flew 47 h, and prompted development of a real turboprop.

This matured as the RB.39 Clyde. Hooker calculated that a pr of at least 6 was desirable, and after studying 2-stage centrifugal compressors picked a new arrangement: an axial followed by a centrifugal. Thanks to Metrovick, Dr Smith's existing F.2 spool was offered, and the second stage was a single-sided impeller scaled up from the Merlin 46 supercharger. The axial was derated to 9 stages (pr, 2.65) running at 6,000 rpm, while the 2.35-pr centrifugal ran at 10,800 rpm, so for the first time the answer was a 2-spool engine, with separate 1-stage turbines driving the HP compressor and (via gearing) the LP axial spool and propeller. The HP spool drove all accessories and was all that had to be cranked for starting. Combustion was in 11 skewed chambers. The Clyde first ran as a complete engine on 1 August 1945. After a hiccup due to an error in matching the centrifugal and axial almost every achieved figure was a little better than predicted. The first run was at 2,000 shp, and altogether 9 Clydes ran, No 9 demonstrating 4,200 shp (4,543 ehp). Had the overloaded LP turbine been replaced by a 2-stage turbine considerably higher power and efficiency would have been achieved. The Clyde proved to be an outstanding and reliable engine, and, following severe tests to clear it for 500 mph at full throttle at sea level, an order was placed for 100 engines for the Wyvern. Hives refused this; he claimed the Avon turbojet would be 'the Merlin of the future' and that turboprops would soon be superseded. In view of the subsequent history of the Python and Proteus this was a great pity.

Herriot managed the Clyde development almost alone at Barnoldswick. Hives had by this time decided to close what he scathingly called 'Hooker's bloody garage' at the remote northern site, and bring all gas turbine work to company HQ at Derby. With

piston engines apparently being phased out, he was afraid of the jet tail at Barnoldswick wagging the giant dog. Even the final development of the centrifugal jet was done at Derby. This was the RB.44 Tay, a Nene redesigned to 115 lb/s and with many detail improvements. Initially rated at 6,250 lb, it found no application except in licensed forms by Pratt & Whitney and Hispano-Suiza, another failure by Britain to use a world-beating engine.

Part of the trouble lay in the belief, fostered during Whittle's early struggle by such experts as Griffith and Constant, that not only were axials better than centrifugals, but that the latter were a crude idea put forward by Whittle that would soon be rendered obsolete. Nothing could have been further from the truth, but in 1944 the centrifugal was limited by the strength of the available aluminium alloys to straight radial vanes and tip speeds around 1,500 ft/s. This meant about 80 per cent efficiency at pr of 4. Today titanium impellers with curved vanes and tip speeds exceeding 1,800 ft/s achieve 84 per cent at pr of 8.4 or more, which is far beyond what any axial could do in 1944! But RR decided to abandon the classic Whittle formula, and in early 1945 Hives ordered Griffith to start on the AJ.65 (axial jet, 6,500 lb) a simple engine by contrast to the CR complexities. After doing preliminary designs in June 1945 it was handed to Hooker, then still at Barnoldswick. From the start there was trouble, beginning when Lombard calculated that Griffith's optimistic weight would be exceeded by 50 per cent. When the prototype was started in spring 1946 it refused to accelerate, broke its first-stage blades and could hardly reach 5,000 lb thrust. Sadly, this triggered deep disharmony between Hives and Hooker which led to the latter's departure at the end of 1948.

Though RA.1 prototypes had only 8 or 10 stages, the AJ.65 was designed with a 12-stage compressor achieving 20°C per stage and pr of 6.5 overall at 85

Though it looked a bit like two engines joined together, the Clyde 2-spool turboprop performed brilliantly and reliably. The author always regretted Hives' refusal of the production contract.

per cent efficiency, handling 120 lb/s. Eight Lucas chambers then led to a large 1-stage turbine. Gradually the compressor was made to behave, though it needed variable inlet guide vanes and blow-off valves linked with an automatic starting and acceleration control. Early RA.2 engines weighed 2,550 lb and gave 5,800 lb, but in 1949 a new 2-stage turbine increased power and saved 300 lb in weight in the Avon RA.3. The first 100-series production engines were delivered in 1950 at the RA.3 rating of 6,500 lb and a weight of 2,240 lb. Almost all had a **cartridge starter and direct compressor bleed for cabin pressurization**. The later 100-series at RA.7 rating of 7,500 lb (later 8,050) had full anti-icing and other improvements, and were made by Derby, Bristol, Napier, Standard Motors and CAC

Top *In ironic contrast to the Clyde, the AJ.65 Avon looked fine but performed abysmally. This early RA.1 at least has the actuator above the compressor driving the variable inlet vanes, and a row of 3 blow-off valves at the upstream end, but it retains the massive 1-stage turbine.*

Above *The same size as the RA.1, but a bit heavier, this Avon 300-series with much greater airflow and air-cooled blades gave 12,690 lb dry, or 17,110 lb with afterburner. It could have saved Westinghouse.*

(Australia). Hiccups still occurred, such as the discovery that when an Avon-Hunter fired its guns the engine stalled, but the Avon gradually did fulfil Hives' hopes that it would be the 'Merlin of the jet age'.

It really became competitive when in December

1952 it was redesigned as the 200-series with a much better compressor whose first 4 stages followed the aerodynamic design of the Sapphire, and a cannular combustor. The first 200-series engines had a 15-stage compressor of 150 lb/s, with pr 7.45, and were type-tested in April 1953 at the RA.14 rating of 9,500 lb with sfc reduced from 0.92 to 0.84. This thrust was the same as for the 100-series RA.7R with afterburner and primitive twin-eyelid nozzle. In July 1956 the RA.24 of 11,250 lb was type tested, with the first production air-cooled blades. For the Comet and Caravelle the civil RA.29 was produced, with a zero-stage giving 10,250 lb at sfc of 0.786, and the RA.29/6 added a '00-stage' and 3-stage turbine giving 12,600 lb at the same sfc. Fighter Avons culminated in the RB.146 300-series for Lightnings and produced by

Forty years from the prototype, this Dart 551 was in 1985 installed in an F27. Rolls-Royce now regret having underestimated the importance of the 'Dart replacement' market.

Svenska Flygmotor as the RM6 for Drakens, with a zero-stage (airflow 170 lb/s, pr 8.43) and an efficient afterburner and multi-flap nozzle giving 17,110 lb thrust. In 1959 an aft-fan Avon was tested outdoors at Hucknall, recording world-record low sfc.

In 1946 Lionel Haworth designed the RB.53 Dart, another Derby engine with tandem centrifugal impellers in close series. Though Griffon and Eagle supercharger aerodynamics helped, getting the air from the first stage into the eye of the second in a compact engine was not simple. The 7 skewed chambers led to a 2-stage turbine, driving both compressors and the compound helical propeller gearbox in the centre of the annular inlet. The Dart flew at 890 shp in the nose of a Lancaster in October 1947, and went into production for the Viscount in 1952 at RDa.3 rating of 1,400 shp. The RDa.6 raised this to 1,600 shp, and the RDa.7 with 3-stage turbine to 1,800 shp, airflow being 21.5 lb/s and pr 5.4. Later Darts were rated up to 3,245 ehp, and several marks are still in production with over 7,100 engines having flown some 114 million hours.

In 1953 effort began to be applied to jet VTOL, resulting in the TMR 'Flying Bedstead' powered by 2 Nenes, and also in the RB.93 Soar, first of a series of light and simple turbojets. The Soar was intended mainly for RPV propulsion, with thrust and weight of 1,810 and 267 lb, but it led to the RB.108 lift jet, with corresponding figures of 2,340 lb (with 5.9 per cent control bleed) and 270 lb. This led to the RB.145 used in supersonic V/STOLs for both lift and thrust, rated at 2,750 lb dry or 3,650 lb with afterburner. The prospect by 1961 of thousands of military and civil V/STOLs, each with up to 36 lift engines, was enough to launch massive sales efforts, but the market evaporated. The final production lightweight jet was the RB.162, first run in January 1962. This marked a total breakaway from established practice, with engine length being reduced near to the diameter, and virtual replacement of exotic metals by glass-fibre and aluminium. Lift versions typically had 86 lb/s airflow and gave 4,200 lb for a weight of 280 lb, while for the Trident 3 the RB.162-86 was developed as a long-life take-off booster. Schemes were drawn for attractive lift turbofans of very high bypass ratio, and with diameter much greater than the length.

Following prolonged Griffith studies, government funding was obtained in 1952 for the first bypass jet, the RB.80 Conway. Had the turbofan concept been coined this engine might have had a useful bypass ratio of at least unity, but it was viewed as a 2-spool turbojet with a slightly oversized LP spool, and accordingly the bypass ratio was established at only 0.3, insufficient to make much difference to sfc or noise. The engine ran in 1953 and flew in a pod under an Ashton at RCo.2 rating of 9,250 lb in 1954. Subse-

quent ratings were 11,500 and 13,000 lb for the V.1000, followed by 17,250 lb for the Victor B.2, 17,500 lb for the 707 and DC-8, 20,600 lb for the Victor and 21,800 lb for the Super VC10. The final Mk 550 engine had a zero-stage on the 7-stage LP compressor, handling 375 lb/s, 9-stage HP, 10-tube cannular chamber, 1-stage air-cooled HP and 2-stage LP turbines, weight being 5,101 lb.

Haworth spent 1954 designing the RB.109, a second-generation turboprop to take over at 2,500 hp where the Dart left off. Run in April 1955 and named Tyne, it far exceeded expectations and was quickly type-tested at 4,220 shp (4,690 ehp) at 15,250 rpm with airflow of 41 lb/s and pressure ratio of 13.5. It had a 6-stage LP compressor driven by a 3-stage turbine with the shaft extended forwards to the compound epicyclic reduction gearbox centred in the annular inlet, a 9-stage HP spool driven by a 1-stage

Top *Last of the Conways, a Mk 550 (RCo.43) of 21.825 lb rating, flies RAF VC10s with high reliability, outer engines having the reverser seen here. The only criticism of the Conway was timidity over choice of bypass ratio.*

Above *The first 'two-spool' turboprop, the RB.109 Tyne proved a real winner. Unfortunately it hit the market just as turboprops seemed to be being replaced by jets, and it never found a worthwhile application. Today Tynes in the 8,000-hp class may be used for propfan tests.*

air-cooled turbine, and 10-tube cannular combustor. Today airflow is 46.5 lb/s and pr 14. Tynes are rated at 6,100 ehp, weight remaining just over 2,000 lb, and a few are still being made by an RR/SNECMA/MTU/FN consortium.

In 1957 design began on the RB.140, a totally new bypass jet of 8,000 lb thrust. This became the 14,000-

lb RB.141/142 Medway, for the DH.121 and, with a switch-in deflector to vector the thrust, the HS.681. Seven development engines ran well. Foolishly BEA demanded that the former aircraft be made smaller; the Medway, which would also have powered the Saab Viggen, was therefore scaled down to 200 lb/s to become the RB.163, of 9,850 lb. Designed by Freddie Morley, this was a neat 2-spool engine with 4-stage (later 5) LP, 12-stage HP, 10-tube cannular combustor and 2-stage HP and LP turbines, with the first HP stage air-cooled. Named Spey, it gained an important civil and military market, with ratings around 12,550 lb (wet) or 20,500 lb with afterburner. Previously named Spey Junior, the RB.183 powers the F28; it has a 4-stage LP, reducing pr from around 21 to 15 but raising bypass ratio from 0.64 to 1. Almost 20

years were spent studying Spey derivatives before deciding in 1983 to retain the HP spool but add a new LP with single-stage fan (with advanced wide-chord blades rotating with a 3-stage compressor), driven by a new 3-stage LP turbine. Named Tay, this engine

Below *Previously called Spey Junior, the RB.183 Mk 555 is a simplified engine which powers the F28. Rated at 9,850 lb, it weighs 2,250 lb.*

Bottom *Rated at 20,515 lb with afterburner, the Spey version for Mach-2 fighters was produced as the Mk 202 for RAF Phantoms and the Mk 203 (with fast afterburner light-up for carrier overshoots) for the Royal Navy. The People's Republic of China took a licence, built a mighty factory, delivered an engine that sailed through its qualification tests, and then dropped the idea!*

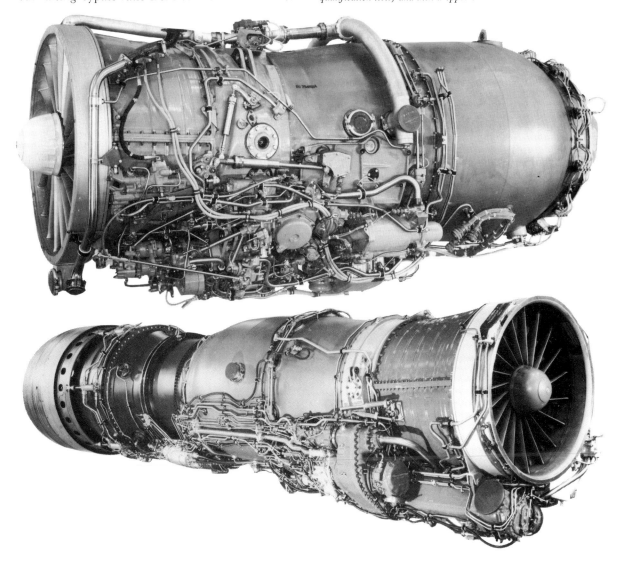

Left *First run in August 1984, the Tay was by 1985 giving promise of being another winner, with performance well beyond prediction. The first five Tays romped through the first year of testing in a way that can fairly be called unprecedented.*

Below right *Derby has come a long way since the ulcer-making first years of the RB.211. These are the latest 524D4-upgrades, which set you back about £5 million each.*

has a bypass ratio of 3 and begins life at 13,550 lb, with the promise of becoming a best-seller.

One major puzzle is why the Spey bypass ratio was pitched so low; a contributory factor was over-estimation of nacelle drag at higher ratios. A greater puzzle is why, having discovered the engine needed to be refanned in 1966, the go-ahead was delayed for almost two decades! A Spey derivative was produced by Allison as the TF41. More distant relatives were developed from 1960 in collaboration with MAN Turbo (later MTU) of West Germany for V/STOL fighters. The RB.153 was scaled down from the Spey figure of just over 200 lb/s to 121, giving a very compact engine rated at 6,850 lb or 11,645 lb with afterburner. The latter incorporated skewed rotating joints so that the variable nozzle could be rotated down through 90°. In contrast the RB.193 had Pegasus-type pairs of hot and cold nozzles. Though it had the same airflow as the Spey it was a new design with 3-stage fan and 2-stage IP compressor driven by a 3-stage LP turbine, and counter-rotating 6-stage HP spool driven by a 1-stage air-cooled turbine; weight was 1,742 lb bare and thrust 10,163 lb.

In 1961 competition from high-ratio turbofans spurred long-term studies which embraced large 3-shaft engines, which appeared to offer advantages in flexible operation, reduced numbers of parts, rigidity, elimination of variable stators and very low performance deterioration in service. In 1967 the 2-spool RB.178 was tested as a research tool, and in September of that year the definitive 3-shaft RB.207 was selected at 50,000-60,000 lb to power the A300. In March 1968 strenuous sales efforts succeeded in getting the smaller RB.211 of 40,600 lb chosen as launch engine for the L-1011 TriStar. With initial launch orders for 144 aircraft this seemed far more important than the nebulous Airbus, and the latter's reduction in size to A300B standard was used as an

excuse for the British government to withdraw from the consortium.

Sir Denning Pearson at Derby was perfectly happy to ignore the A300B, and ordered all Derby resources harnessed to the RB.211. By far the company's biggest project, in all senses, its 5 modules included a single-stage 89-in fan with 25 blades of Hyfil (carbon-fibre composite), which seemed an ideal application for this revolutionary material which made possible supersonic lenticular-profile blades of wide chord and extremely low weight, the weight-saving being multiplied many times in enabling other parts of the engine and airframe to be made lighter. This fan was driven by a 3-stage LP turbine. The 7-stage IP spool included glass-fibre in its construction, and was driven by a 1-stage turbine. The 6-stage HP spool was driven by a 1-stage air-cooled turbine. All turbine blades were of the traditional wrought type. By this time RR had ceased to subcontract combustion systems to Lucas, and Derby designed the fully annular chamber with 18 atomizing burners. Weight was to be 6,353 lb, a figure that proved as optimistic as the price.

For the first time, and bereft of Lombard who died suddenly in 1967, the Derby team had bitten off more than it could chew. The Hyfil blades proved unable to meet the requirements for birdstrike and erosion, and had to be replaced by 33 narrower titanium blades, with part-span snubbers, at a stroke losing what had been sold (some say 'oversold') as one of the chief technical advantages. The turbines proved to have pathetic performance, and the forged air-cooled blades for 1,250°C proved incredibly difficult and costly to make. Bristol, bought in 1966, might have made them far more easily using investment casting but was not asked, nor was it asked to contribute vaporizing burners. There were also multiple mechanical problems, notably affecting bearings and seals, which for an engine of 27 pr are crucial to the attainment of acceptable efficiency. It added up to an engine that was nowhere near guaranteed performance, uncertifiable, and thus unsaleable. With insufficient money coming in, and millions a week going out, RR declared itself bankrupt on 4 February 1971. When the government formed Rolls-Royce (1971) Ltd 19 days later it excluded the RB.211, but after prolonged negotiation a conditional fresh contract with Lockheed was signed on 11 May.

Previously, in 1970, Dr S.G. Hooker had been recalled from retirement to study the RB.211, and as a result a redesigned RB.211-22 went on test the day before the bankruptcy. It improved thrust on Standard Day bench test from 34,000 lb at 1,167°C to 39,340 at 1,227°, followed by 41,500 lb with modified nozzle guide vanes and 43,500 lb with cast blades. In February 1973 the -22B was certificated at 42,000 lb to 28.9°C, the weight being 9,195 lb. Airflow is 1,380 lb/s and bypass ratio 5. Alec Harvey-Bailey is convinced that 'had it not been for the 3-shaft modular concept, which enabled sick engines to be dealt with rapidly, we should not have survived the introductory years'. Hooker had meanwhile completely redesigned the engine in order to compete in the 50,000-lb market, and the resulting -524 ran on 1 October 1973 after two years waiting for permission to build it. It has a new fan of unchanged diameter but handling 1,550 lb/s, with bypass ratio 4.4, improved compressors, a new HP turbine, and a larger and simpler jetpipe. Today a series of improved -524 engines have been certificated at ratings of 50,000 to 54,000 lb with progressively lower sfc, all weighing about 9,800 lb. They have the best performance retention of all big fan engines.

In 1974 RR(71) studied the market for engines in the 20,000-30,000-lb class and offered a scaled-down RB.211-524 known as the RB.235. This was replaced in 1979 by the RB.211-535, later redesignated Rolls-Royce 535C to de-emphasize kinship with the RB.211. It has an HP module based on the -22B, a 6-stage IP compressor without variable stators, and an advanced -524 fan scaled down to 18 per cent less airflow. Weighing 7,294 lb, the 535C is conservatively rated at 37,400 lb, and has proved the most reliable engine of all time. From it has been derived the 535E4, with a new fan with 22 wide but light snubberless blades formed from titanium skins on a bonded honeycomb core, and revised compressor and turbine blading to improve efficiency. Certificated at 45,000 lb, it has outstanding fuel efficiency and promises to beat even the 535C for reliability. The new fan is probably RR's greatest single lead over the competition, and is the result of 10 years' development.

In 1973 various small turbofans were studied, and an RB.401-06 was run in 1975, but no go-ahead for an engine in this class has been forthcoming. At Leavesden, formerly DH Engines, the Small Engine

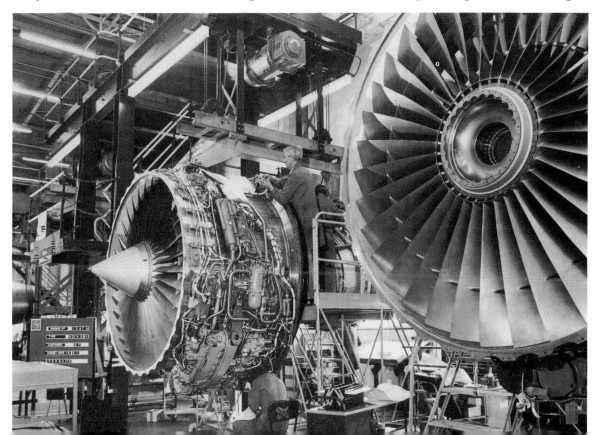

a 2-stage fan driven by a 1-stage turbine, 5-stage compressor (95 lb/s, overall pr 11) driven by a 1-stage air-cooled turbine, annular combustor and, in most applications, an afterburner. France handles the compressor, casings, external piping and (assigned to SNECMA) afterburner; RR handles the rest. Ratings with afterburner vary from 7,305 to 8,400 lb; about 1,800 have been delivered, plus some 500 licensed to Japan, India and Finland.

Since 1978 the two partners have studied engines for helicopters, and as this book was written the first RTM.322 began bench testing. Rated initially at up to 2,308 shp, the 322 has a 3-stage axial plus 1-centrifugal compressor, annular reverse-flow

Below *By mid-1985 eight Rolls-Royce Turboméca RTM 322 turboshaft engines were running, showing clear promise of later ratings up to 3,000 shp. This is the biggest of the new engines covered by a three-nation agreement, the third party being MTU of West Germany, and it has been licensed to Pratt & Whitney.*

Bottom *Despite the complexity of the dressing (the external pipework and accessories) the Rolls-Royce Turboméca Adour is one of the simplest engines, and these afterburning examples are only 117 in long.*

Above *Former astronaut Frank Borman, President of Eastern Airlines, calls the Rolls-Royce 535C 'The finest airline engine in the world'. The Derby engineers might reply 'You ain't seen nothin' yet, here's our new 535E4'. The E4 entered service with Eastern in 1984.*

Division licences an Anglicized T58 as the Gnome, and has developed the BS.360 into the Gem. A 3-spool turboshaft, the Gem has 7 modules, one comprising the 4-stage LP compressor and 1-stage shrouded turbine, another the centrifugal HP compressor and 1-stage air-cooled HP turbine, another the annular reverse-flow combustor, and another the 2-stage power turbine and jetpipe. The power turbine can give direct drive at 27,000 rpm or a 6,000 rpm output from the double-helical gearbox centred in the air inlet. Pr is typically 12, airflow 7 lb/s, weight 320 lb, and output 900 to 1,200 hp.

In 1986 the company runs the first XG-40, an augmented turbofan for future fighters, with thrust/weight ratio exceeding 11. It is a 2-shaft engine in the 22,000-lb class.

The Adour and RTM.322 are described under RRTI, the RB.199 under Turbo-Union and the V2500 under IAE.

RRTI (FRANCE/UNITED KINGDOM)

Rolls-Royce Turboméca Ltd was formed by the two companies in June 1966 to manage the programme for the Adour turbofan, initially for the Anglo-French Jaguar. The Adour was conservatively designed with

combustor and 2-stage turbine with air-cooled first stage and uncooled single-crystal second stage. The separate 2-stage power turbine drives to front or rear.

Royal Aircraft Factory (UNITED KINGDOM)

In 1912 HM Balloon Factory became the RAF (Royal Aircraft Factory), and by November of that year had produced a conventional water-cooled 6-in-line, Colonel Mervyn O'Gorman's chief engineer being Major F.M. Green. Other engineers who became famous included G.S. Wilkinson, Jimmy Ellor, Sam D. Heron, Professor A.H. Gibson and A.A. Griffith. In early 1913 the RAF.1A was completed, based on the V8 Renault but with larger cylinders (100 × 140 mm, 8.8 litres), aluminium pistons, and various other changes including a scoop air inlet with no cooling fan except in pusher installations. The heads were not detachable from the cast-iron cylinders but had an overhead exhaust valve and an inlet valve in a detachable pocket. Oil was carried up from the sump by a light flywheel, to be scraped off at the top and allowed to feed by gravity to bearings and gears. As in the Renault, a very rich mixture was used, to help avoid overheating. Rating was 92 hp at 1,600 rpm.

In late 1915 the RAF.1B introduced 105 mm bore cylinders (9.7 litres), with differences to the cooling baffles and oil system, running at 1,800 rpm to give 115 hp; it was mass-produced by car firms. The RAF.2 was a 120-hp 9-cylinder radial designed in

Above left *The RAF.1A air-cooled V-8 was made in very large numbers, mainly for BEs. It was rated at 90 hp.*

Above *Close up of the later RAF.4A, a V-12 also made in great quantities. It has its air cooling scoop and baffles fitted.*

October 1913. Various water-cooled engines were built, including the 200-hp RAF.3 of September 1914, but the other mass-produced engine was the RAF.4, a V-12 first run in December 1914. The cylinder banks were not at 90° but at 60°. With RAF.1A-size cylinders the power was initially 90 hp, a poor figure for 13.19 litres even with compression ratio of 4.5 and over-rich mixture. There were two carburettors and two 6-cylinder BTH magnetos, lubrication again being by flywheel pickup, weight being 605 lb.

In 1916 the improved RAF.4A, weighing 637 lb, was rated at 140 hp but at 1,800 rpm actually developed 160, and made up the bulk of the 7,000 Factory engines made by car firms in 1915-17. By 1916 Gibson and Heron had carried out the first systematic research into air-cooled cylinders performed anywhere, developed gear-driven superchargers, and begun work on aluminium cylinders. In early 1917 the RAF.4D (100 mm bore) and RAF.4E (105 mm bore) introduced vastly superior cast aluminium cylinders with integral oversize spherical heads incorporating brackets for the overhead valve gear. Open-ended steel liners were pressed into the hot cylinder. With compression raised to 4.7 these much better engines gave 220 and 240 hp respectively at

2,200 rpm, for a weight of 670 lb. Service rating was 196 hp at 1,800 rpm, but they never went into production. The RAF.5 was a fan-cooled pusher version. The RAF.8 of September 1916 was a totally new 14-cylinder radial using cylinders designed with help from the RAF.4D/E. It became the Armstrong Siddeley Jaguar. In 1917-24 the Factory, renamed the Royal Aircraft Establishment, pioneered turbochargers, but developed no more engines. Griffith's axial-compressor research is discussed under Armstrong Siddeley and Metrovick.

S

Salmson (FRANCE)

Emile Salmson's modest company decided to make aero engines in late 1911. Early in 1912 the Billancourt (Paris) works produced the first of a long series of unceasingly improved static radials of novel design. At first they were water-cooled and used the patented (Swiss Canton-Unné) system in which all conrods drove a cage revolving on the crankpin on epicyclic gears. They had a separate cam ring for each pair (one inlet, one exhaust) of valves for each cylinder, with short stems and hairpin springs to reduce diameter. Some engines were mounted with the crankshaft vertical, driving through a bevel gear. In some applications two engines faced outwards from the fuselage to drive left/right propellers by transverse shafts and outboard bevel gears. The initial families had 120 × 140 mm cylinders, the 7-cylinder giving 90 hp and the 9-cylinder 110 to 140 hp. By 1917 the conventional master rod was being introduced on the important Z9 series, usually just called the Salmson 260-hp, which had cylinders 125 × 170 mm.

Despite the profusion of types and miniscule design staff the Salmsons were reliable; Handley Page picked a 200-hp model for his single-engined Transatlantic L/200 of 1913. In the 1920s there was an attempt at rationalization, though the range included the 9-cylinder AD9 with bore/stroke only 70 × 86 mm, a size now returning for microlights! From 1920 all Salmson engines were air-cooled, with few unusual features. In 1946-7 the firm tried to market the Argus As 10C as the (240 hp) 8AS, as well as the pre-war (45 hp) 9ADB, (90 hp) 5AQ, (175 hp) 9ND and (230 hp) 9ABC, all radials. Liquidation followed in 1951.

As odd inside as outside, the 9-cylinder Salmson Canton-Unné of December 1912 was a water-cooled radial, in this instance with bevel drive. By 1917 Salmson was building more conventional radials, biggest being the 530-hp 18Z of February 1918, with pairs of 9Z cylinders one behind the other.

Above *London's Science Museum has this fine Salmson Canton-Unné water-cooled radial, typical of thousands made in 1914-17. These were notably smooth-running and reliable engines.*

Above right *In 1927-59 A.D. Shvetsov's M-11 was virtually the standard engine for Soviet trainers and lightplanes. Deliveries exceeded 100,000.*

Right *Qualified in 1937 at 840 hp as the M-62, Shvetsov's ASh-62 has been made in useful numbers, the An-2 biplane alone taking well over 20,000. Production was transferred to Poland in 1952, this being a PZL-built ASz-62IR, of 1,000 hp.*

Shvetsov (SOVIET UNION)

Arkadiya Dmitriyevich Shvetsov was the first aero-engine designer to establish himself in the Soviet Union; in 1930 he was a founder of TsIAM, the central institute for aero engines. His firstborn was the unimpressive M-8 radial of 1925 (previous numbers were assigned to foreign engines). From this he developed the M-11 radial, one of the world's classic engines with over 100,000 and probably over 130,000 built in 1927-59. It had 5 cylinders 125 × 140 mm, 8.6 litres, and was qualified in 1928 at 100 hp at 1,590 rpm using any available motor spirit (later handbooks specified 59 octane). Subsequent versions gave 115, 145, 160, 165 and 200 hp at up to 1,980 rpm using better fuel. The M-12 was a 1930 development of 190 hp. The M-15 of 1929, a 9-cylinder engine of 450 hp, did not go into production.

The M-21 was a 200-hp 7-cylinder from which Kossov derived the MG-21. Shvetsov's bureau

About 70,000 ASh-82 14-cylinder engines were built, this example being a post-war ASh-82T airline engine of 1,900 hp. In 1941-42 Lavochkin and TsIAM achieved an outstanding installation in the La-5 fighter.

managed the development of the licensed Jupiter into M-22 variants, and used M-15 cylinders in the M-26 of 300 hp based on the Bristol Titan. Shvetsov also played a managerial role in developing the Cyclone (M-25) into the M-62 by fitting a two-speed supercharger, improved induction system and other changes. In 1941 this was redesignated ASh-62 in accord with the new General Constructor scheme. Including the M-63 of 1939 rated at 1,100 hp, total production by 1985 exceeds 67,000, Polish designation being ASz-62 and Chinese, HS-5. The ASh-71 to -73 were 18-cylinder engines using M-63 cylinders, rated at 1,700 hp in 1941 and up to 2,650 hp in the 73FN of 1944.

Most important of all Soviet radials was the ASh-82, a 14-cylinder 2-row engine with M-62 cylinders with stroke reduced to 155 mm (bore remaining 155.5), giving 41.2 litres in a compact 1,259 mm diameter, which was the basis of the world's first good air-cooled fighter installations apart from the Fw 190. Qualified in 1940 at 1,250 hp on 87-grade fuel, the ASh-82 was developed in 22 versions up to 2,000 hp. About 70,000 were produced, some still being in use in Il-14s and Mi-4 helicopters, the latter having a 25° inclined installation with direct drive and cooling fan. Few ASh-83 1,900-hp engines, or 1,500 hp 18-cylinder ASh-90s (1941) were made, and the ASh-2 of 1950 (two ASh-82s in tandem giving a 3,300-hp unit) did not go into production. Substantial numbers were made of the 7-cylinder ASh-21, in effect half an ASh-82, qualified in 1947 at 700 hp and developed to 760 hp for aeroplanes and helicopters.

Siemens (GERMANY)
See Bramo.

SNECMA (FRANCE)
On 29 August 1945 the Gnome-et-Rhône company was given the title Société Nationale d'Etudes et de Construction de Moteurs d'Aviation, on its enforced nationalization. In 1946 SNECMA took over the previously sequestrated Renault company, as well as SECM (Lorraine) and the GEHL oil-engine works. In 1968 SNECMA also took over Hispano-Suiza's engine business and licences.

At the start the production engines were the GR14N, R and U, Regnier 4L and Renault (Argus-derived) 12S. Under Henri Desbruères work went ahead in 1947 on simple pulsejets which had been worked on from 1943 by Ingenieur en Chef de l'Air Raymond Marchal (by 1946 SNECMA technical director). These were purely resonant ducts without valves or other moving parts. The *Escopette* (carbine) weighed 4.8 kg and gave 10 kg thrust; the *Ecrevisse* (crayfish), bent back on itself in a U, with inlet facing aft, came in various sizes with thrusts from 10 to 150 kg. SNECMA's first gas turbine was the TB.1000 turboprop. Design began in February 1946 and settled as a simple single-shaft engine with 9-stage compressor, 6 combustion chambers and 2-stage turbine with straight 0.109 spur gear to the propeller. The 1948 prototype gave 1,300 shp, being followed in 1951 by the TB.1000A of 1,760 shp (2,000 ehp), weight being 450 kg. It never flew.

Far more important was the Germanic turbojet designed in a small office at Rickenbach, Switzerland, by Dr H. Oestrich, former BMW 003 chief engineer, and several key colleagues, who escaped there in May 1945. They had made contact with the French, and adopted a French stance by naming themselves the Atelier Technique Aéronautique Rickenbach, their engine becoming the Atar in consequence. Design was complete in October 1945, and two months later the Ministère de l'Air awarded a development contract on the understanding all manufacture would be done in France. A month later the ATAR was closed and the team relocated at Decize (Nièvre) with the delightful name 'Aéroplanes G. Voisin, Groupe O'. In June 1946 the drawings for the Atar 101V prototype were sent to SNECMA. It was made at the latter's Usines Kellermann and Gennevilliers, and assembled in March 1948 at the Melun Villaroche centre, then almost a virgin site. It was first run a month later on a lash-up bench of steel girders on four wheels. The 101V had a 7-stage compressor (pr 4.2 at 8,050 rpm), annular combustor with 20 burners, 1-stage turbine with 53 wrapped-sheet air-cooled blades, and nozzle with a large central bullet translated in/out by a hydraulic ram on the centreline. The entire engine was made of ordinary commercial steels, nothing else being available. By 1949 improved 101A engines were flying in a

Above *The very first Atar, the 101 V1, at Decize in 1948.*

Below *One of the final Atar versions was the 8K-50, used in the Super Etendard. It is one of the few models without afterburner.*

Marauder and on a pylon above a Languedoc. By this time SNECMA was not only chaotic but near collapse, with vast overheads and payroll (16,950 in 1947) but practically no sales apart from licensed Hercules piston engines. Workforce was slashed to 6,600, and in June 1950 Oestrich's team was formally taken over, the Atar becoming a SNECMA property.

The 101B introduced solid Nimonic turbine blades with correct twist, as well as improved combustor airflow, and a thrust of 2,350 kg was established in Meteor *RA491* (previously Avon-engined) in late 1949. The B2 had extra stator blades, raising pr to 4.4. The C achieved production for the Mystère IIC at 2,745 kg, with speed raised to 8,500 rpm, Air Equipement (Rotax) starter in the nose bullet, revised combustor, and nozzle bullet separated from the tailcone. The D reverted to 8,300 rpm, but had a larger turbine and new nozzle with the bullet replaced by upper/lower eyelid flaps. The E introduced a zero-stage handling 60 kg/s at 8,400 rpm with pr 4.8; the enlarged turbine diameter was carried forward through the combustor and in the E4 (3,700 kg rating, 880 kg weight) the nozzle eyelids were improved. The F was a D with afterburner, tested in 1952 with much larger nozzle eyelids and qualified in 1954 at 3,800 kg. The G was an afterburning E, the G2 and G3 giving 4,310 kg and the G4 4,700 kg.

In 1954 design went ahead on the new-generation Atar 8. Amazingly this modest improvement has led to the Atar, despite its now totally obsolete design, still being in volume production—SNECMA having concentrated solely on the supersonic fighter market. The 9-stage compressor handles 68 kg/s at pr of 5.5 at 8,400 rpm, driven by a 2-stage turbine of reduced diameter. Construction was improved throughout, rating of the 08B3 being 4,310 kg. The Atar 9 (originally 09) was the 8 with afterburner, but introduced a new compressor with a cylindrical drum formed from centreless discs and a casing of ZRE-1 magnesium alloy. The afterburner is enlarged and

improved, 9B3 rating being 4,170 kg dry and 5,880 maximum. The 9C introduced a further enlarged afterburner with 18 nozzle flaps, as well as a Microturbo gas-turbine starter which established a new aero-engine company! Maximum thrust increased to 6,200 kg, and at over Mach 1.4 an over-speed to 8,700 rpm was permitted. In the 09D the compressor was largely titanium, for sustained Mach-2 flight, and this led to the 9K with improved combustor, turbine cooling and afterburner, ratings being 4,610/6,570 kg in the K-6 and K-10 (Mirage IVA). By 1967 this led to the 9K-50, with improved compressor blading (pr, 6.5), a new turbine with cast and coated blades, and many other changes. Weighing 1,587 kg, ratings are 5,015/7,200 kg at 8,700 rpm. Without afterburner the corresponding 8K-50 delivers 5,000 kg. Total Atar production exceeds 5,000.

In June 1951 design began on the R.104 Vulcain, basically an early Atar enlarged to an airflow of 82 kg/s. First tests on 21 May 1952 reached 4,500 kg, and a little flying was done in a pod under an Armagnac in 1954, but the Vulcain was abandoned at 6,000 kg in January 1955. The R.105 Vesta, of December 1953 design, was a Vulcain scaled down to 24 kg/s at 12,900 rpm and intended to have an afterburner in twin-engined light fighters. It ran in December 1954 at 1,200 kg but was dropped before afterburning trials began. To fill the gap and replace the seemingly outdated Atar the Super Atar was designed in 1956-58 as an advanced single-shaft variable-stator engine for 9,000 kg thrust with afterburner. But after discussion with Dassault and

the government it was decided to import foreign technology; Pratt & Whitney was chosen and licences obtained for all that company's piston and turbine engines in exchange for 10 per cent of SNECMA's stock and membership of the board of directors (unusual for a nationalized concern). The June 1959 deal brought SNECMA immediate military and airline overhaul business.

It was also intended to lead to production of a French augmented turbofan derived from the JTF10A. In November 1961 the go-ahead came on the TF-106, derived from the JTF10A-20 (TF30 variant) with an advanced SNECMA afterburner. To gain experience SNECMA made TF-104 engines based on the JTF10A-2. In October 1963 a TF-104 stalled on its first take-off in the Mirage IIIT (an aircraft much bigger than other Mirage IIIs) but aborted safely. It finally flew on 4 June 1964, the first flight by an augmented turbofan. (P&W were later to suffer severe stall problems in the F-111.) Five TF-104Bs were followed by 13 TF-106 engines, of advanced supersonic design, first run in October 1963, flown under the Armagnac in early 1964 and in the Mirage IIIT on 25 January 1965, reaching Mach 2.05 on 26 November 1965 (before any other augmented turbofan had reached Mach 1). The last 106A3 was rated at 7,900 kg.

In early 1964 the deep problems with the TF30 caused SNECMA to consider switching to the Spey, but the US firm's hold over SNECMA prevented this. Instead in March 1964 SNECMA agreed to develop a less-troubled engine, the TF-306, starting not with the 1959 JTF10A but with the 1964 TF30-P-1. Eight TF30s were supplied by P&W, one flew under the Armagnac in December 1965, and another powered the high-wing Mirage F2 on 12 June 1966. Studies were made for the TF-306 inside a Nord ramjet fighter at Mach 4. The TF-306E powered the Mirage G, but in 1968 it began to appear that 'the P&W connection' had been a costly waste of ten years. The Mirage G was redesigned with two Atar 9K-50s and the Super Atar was looked at again.

In the mid-1960s SNECMA had two further diversions. By far the larger was the M45 Mars, announced at the 1964 Hanover air-show as a totally new range of turbofans for anything from trainers to supersonic fighters and transports. Bristol Siddeley, already a partner on the Concorde Olympus 593,

joined as equal partner on 1 January 1965 to handle the HP spool and combustor, to form a common core for all M45s. The name Mars was dropped, and the thrust bracket widened to 1,500-6,000 kg. The HP spool was that of the BS.116, itself a scaled Olympus 320 (TSR.2). There is little point in describing the 8 subsequent major variants, which included the 45G for the AFVG swing-wing fighter and 45H for the German VFW 614 airliner. Not one found a major application, though the M45H did power the 614 in service at 7,600 lb thrust and was taken over by Rolls-Royce (Parkside works, Coventry) at SNECMA's request in 1976. RR modified one into the 45SD-02 with a large Dowty Rotol variable-pitch fan, tested at Aston Down. The other diversion was the Larzac turbofan for the Alpha Jet trainer. Despite having SNECMA number M49 this was from the start a 50/50 project with Turboméca. A conservatively designed engine, it has a 2-stage fan, 4-stage HP spool (overall pr, 10.6), annular vaporizing combustor and two single-stage turbines. Over 1,200 04-C6 engines were delivered at a rating of 1,345 kg, production being shared with KHD and MTU of West

Right *One of the many M45 versions that fell by the wayside, the M45.G2 turbofan was a joint project with Bristol Siddeley to power the Anglo-French variable-geometry aircraft in July 1966. With bypass ratio of 1.21, it was to be rated at 13,230 lb with afterburner.*

Below *First run in early 1984, SNECMA's M88 is being developed as the engine of the next generation of French fighters. It is politically important to France.*

In production to power the Mirage 2000, the M53 is a single-shaft engine optimized for supersonic flight. This is a current M53-5 rated at 9 tonnes.

Germany. Telédyne CAE marketed it without success from 1973 as the CAE 490.

When the Super Atar was dusted off in 1968 it was given the new designation M53. Later the name was dropped. A continuous-bleed turbojet, it finally ran in February 1970, flew in the right pod of a Caravelle on 18 July 1973 and powered the Mirage F1-M53 in December 1974 (this aircraft was a contender for NATO orders won by the F-16). In March 1978 the Mirage 2000 began its flight trials. The M53 has a 3-stage LP and 5-stage HP spool rotating together. annular combustor and 2-stage turbine. At 10,500 rpm pr is a modest 9.3, and thrust with full afterburner 9,000 kg, the P2 engine of 9,700 kg being due for delivery in 1985; a typical weight is 1,450 kg. To provide a newer and much lighter engine for future

fighters SNECMA has since 1980 designed the M88. This augmented turbofan has variable inlet guide vanes, 3-stage LP, 6-stage HP, annular combustor and 1-stage HP and LP turbines. The first M88 ran in January 1984 and is to be rated at 7,500 kg; weight has been brought down to 900 kg.

SNECMA is a 50/50 partner in CFM, participates in manufacture of the CF6-50 and -80, and is a member of a four-nation team building the RR Tyne.

SOCEMA (FRANCE)

In 1941 a team of engineers in the Unoccupied Zone, under P. Destival, began design of a turboprop called TGA 1 ('Turbo Groupe d'Air', but as a cover said to mean 'Turbo Groupe d'Autorail', the actual contract being placed by the SNCF). Several examples were run from 1943, and in 1945 the TGA 1bis introduced improvements. Both had a 15-stage compressor of 3.6 pr, remarkably good cannular combustor, 4-stage turbine and epicyclic propeller gearbox. Output was to be 3,000 hp (about 2,400 was achieved), weight being 2,100 kg. In 1945, using partly Jumo 004B technology, the TGAR 1008 turbojet was designed, for SNCASE. Weighing 1,250 kg, it gave the required 1,900 kg thrust and was developed by 1949 to 2,000 kg before being abandoned. These were remarkable achievements with no special alloys and no outside help.

Soloviev (SOVIET UNION)

P.A. Soloviev was unknown until the late 1950s, when his bureau was named as source of the D-15 engines fitted to the '201-M' record-breaking aircraft. This aircraft was later identified as a version of the M-4, normally powered by AM-3 (RD-3) engines much less powerful than the D-15 of 13,000 kg rating.

Soloviev's D-20P was the Soviet Union's first production turbofan. The first fan stage has supersonic tips, and the HP spool has automatic bleeds round stages 3 and 4 to avoid blade stall.

Right *The huge Mi-6 helicopter was planned 30 years ago in parallel with P. Soloviev's D-25V engine. The front end is de-iced by hot oil, and the free turbine drives at the rear through the handed jetpipe.*

Right *The D-30 has part-span fan shrouds and tip shrouds on the turbine blades. No parts are common to the larger D-30K family, despite similarity of designation. This Series II engine has a reverser.*

Right *Despite its designation the D-30KU is almost twice as powerful as the earlier D-30, and is quite different in design. This example is fully dressed, plus reverser.*

It is assumed to be a 2-shaft turbojet which probably influenced Soloviev's better-known airline engines. The D-20P is a 2-spool turbofan with 3-stage fan, 8-stage HP, 12-tube cannular combustor, single-stage HP turbine and 2-stage LP. Airflow is 113 kg/s at 8,550 rpm, pr 13, bypass ratio 1, thrust 5,400 kg and weight 1,468 kg. The D-25V is the big turboshaft designed in early 1954 to power the Mi-6. It has a 9-stage compressor (pr, 5.6 at 10,530 rpm), 12-tube cannular combustor, single-stage compressor turbine and 2-stage power turbine driving to the rear. Weight is 1,325 kg (but the helicopter gearbox weighs a formidable 3,200 kg), and output 5,500 hp. The D-25VF has a zero-stage and gives 6,500 hp.

The D-30 is a turbofan derived from the D-20, with 4-stage LP, 10-stage HP, 12-tube cannular combustor and 2-stage HP and LP turbines. Weight is 1,550 kg and at 7,700 LP rpm airflow is 125 kg/s, pr 17.4 and thrust 6,800 kg. Despite its designation the D-30K is a totally different turbofan of much greater

power. It has a 3-stage LP, 11-stage HP, 12-tube cannular combustor, 2-stage air-cooled HP turbine and 4-stage LP turbine. Weight (without the usual reverser) is 2,300 kg, and at 4,730 LP rpm the airflow is 269 kg/s, bypass ratio 2.42, pr 20, and thrust 11,000 kg, other versions giving 12,000 kg.

Stal (SWEDEN)

The Svenska Turbinfabriks AB Ljungström began in 1946 to design a turbojet named Skuten (witch). The firm had been in gas turbines since 1935 and the Skuten ran in early 1949, but did not fly. It had an 8-stage compressor, 7 combustion chambers and single-stage turbine; weight was 780 kg and thrust 1,450 kg. In 1951 the Dovern went on test, flying in 1953 under a Lancaster. This engine had a 9-stage compressor, 9 chambers and single-stage turbine, weight being 1,195 kg, airflow 55 kg/s, pr 5.2 and thrust 3,300 kg at 7,275 rpm. The Avon was picked instead to power the J35.

Above *The STAL Dovern was a plucky attempt at an advanced axial turbojet at a time when such an engine had almost defeated Rolls-Royce. Unfortunately for STAL, Rolls won through.*

Left *Largest of all the Sunbeams was the Sikh III, a whopping geared V12 rated at 820/900 hp. It had bigger cylinders than the 12 other basic designs, but the author believes it never flew.*

Sunbeam (UNITED KINGDOM)

The Sunbeam Motor Car Company of Wolverhampton got off to a meteoric start after its formation in 1910, producing cars of many types including racers which quickly scored major successes. Technical expertise rested in Louis Coatalen, who came from his native France to build 'motors' in the widest sense. In 1913 he designed two aero engines, a 150-hp 8-cylinder and a 200-hp 12-cylinder. Both were water-cooled V-format engines, the bigger being intended for airships. They were of outstandingly modern conception, with cast cylinder blocks, twin carburettors each feeding one block, enclosed overhead valve gear and plenty of aluminium in the construction. Sunbeam added a 100-hp 6-in-line which was described as 'the only British aircraft engine actually in production at the outbreak of war'.

Coatalen had a good reputation, and was the favoured engine supplier to the RNAS until late in the war. But designs proliferated. Bearing in mind that all used similar technology, and most had a deep V-sump looking like a Hispano, it was often hard to see which engine was actually fitted, especially as the arrangement of cowling, radiator(s) and exhaust(s) often differed as many as 17 times in one aircraft type. The following is an abbreviated list: 1914, unnamed 100-hp 6-in-line, V-8 of 225 hp and prototype '300 hp'; 1915, 150-hp Nubian V-8 (often a pusher), prototype 310-hp Cossack V-12, 190-hp Saracen 6-in-line; 1916, 200-hp Afridi, 100-hp Dyak 6-in-line, prototype Arab V-8; 1917, production Arab at 200 hp (but prolonged difficulty with crankcase strength, cylinder attachment and severe vibration, throwing major programmes into disarray), production Cossack as the '310' (320 hp at 2,000 rpm) and '320'

(345 hp at 2,000 rpm), and major production of Maori V-12s as the '240', '250' or Maori II and '275' or Maori III (actually 265 hp at 2,100 rpm); and 1918, 400-hp Matabele (with two Saracen blocks). Sunbeam made little effort to produce post-war aero engines and in 1923 Coatalen returned to France.

Svenska Flygmotor (SWEDEN)
See Volvo.

T

Teledyne CAE (USA)

Continental Motors obtained a licence in 1951 for Turboméca small gas turbines, the Marboré II turbojet being Americanized as the J69. From this stemmed a wide range of other engines. In 1967 CAE (Continental Aviation and Engineering) was formed as a separate division, and like the parent firm this became a division of Teledyne Incorporated in 1969. It was renamed Teledyne CAE, with headquarters at Toldeo, Ohio, concentrating on small turbojets and turbofans for RPVs, missiles and small aircraft.

The J69 began life as the Continental Model 352, rated first at 660 and then 880 lb. Large numbers of J69-9s were built with pr 4, airflow 18 lb/s and rating 920 lb at 22,700 rpm. Smaller Turboméca engines (6.6-7.5 lb/s) produced the Model 140 and 141 compressors and 220 (T51) turboshaft/turboprop. By 1956 Continental was engaged in adding transonic axial compressor stages, following Turboméca's lead but doing its own design. This resulted in the J69-29 (356-7A), J69-41A (356-29A) and YJ69-406 (356-

34A) RPV engines with airflow around 29.8 lb/s, pr about 5.45 and typically 1,920 lb thrust. Also for RPVs the J100 (356-28A) added 2 axial stages, each with replaceable blades (44.9 lb/s, pr 6.3) to give 2,700 lb at 20,700 rpm. A profusion of further engines followed, as well as a licence extension for the Larzac, but most failed to find markets. Work on lightweight lift jets led to the LJ95 (365) with thrust of 5,000 lb and weight less than 250 lb. Mass production engines include the J402-400 (370) for the Harpoon missile, with precision-cast axial and centrifugal stages (9.6 lb/s, pr 5.8) rated at 660 lb at 41,200 rpm, and the J402-700 (372-2) RPV engine (640 lb at 40,400 rpm). There are several newer engines in the 1,000 or 7,000-lb class aimed at future missile/RPV/trainer markets and having no direct kinship with Turboméca engines. In 1982 Teledyne CAE began second-source production of the Williams F107, cruise-missile engine.

Teledyne Continental (USA)
See Continental.

Tumanskii (SOVIET UNION)

Academician Sergei Konstantinovich Tumanskii became famous for superchargers for high-altitude fighters in World War 2 (when he also began a partnership with the MiG bureau that has strengthened with time). He became deputy to Mikulin, easily made the transition to turbines, and on Mikulin's removal in 1956 was appointed General Constructor of the renamed bureau. Since then his engines have been produced in greater numbers than any other 'make', the total easily exceeding 70,000.

The first turbojet of wholly Soviet design to be mass-produced, the AM-5 was designed under Tumanskii in 1950. From it stemmed the AM-9, with 9-stage compressor (pr 7.14), annular combustor and 2-stage turbine. In the chief application (MiG-19/J-6) all accessories are grouped above the engine and an afterburner is fitted. An engine with accessories on the underside, without afterburner, powered the

Designed by S.K. Tumanskii over 30 years ago, the RD-9BF is still in production in China as the WP-6. This Wo-Pen 6 is seen without afterburner.

Above left *Built in enormous quantities, Tumanskii's very attractive R-11F2S-300 is seen here in the Koraput plant of Hindustan Aeronautics, which has made this fighter engine under licence.*

Above *An HAL-built Tumanskii R-25, a particularly neat 2-spool afterburning turbojet. It has a 2-stage afterburner which is said to make it suitable for air combat at high altitudes.*

Yak-25. In 1956 the AM-9 was redesignated RD-9, and seven major variants were produced until 1959, the RD-9BF-811 continuing in production in China as the Shenyang-built WP-6. All variants have a basic diameter of 813 mm, and a typical maximum rating is 3,300 kg.

From this was derived the R-11, first run in early 1956, a slightly larger 2-spool engine which pioneered the overhung first-stage without inlet guide vanes. All models have an excellent annular combustor and two 1-stage turbines, and accessories are usually grouped under the compressor case. The initial series were rated at 3,900 kg dry and 5,100 kg with afterburner with full modulation and multi-flap nozzle. In the R-11-300 of 1959 an enlarged afterburner and new nozzle increased thrust to 5,950 kg. The R-11F of the same year ran at increased rpm with higher temperature, to give 4,300 kg dry and 5,750 with the original afterburner. The FS and F2S have large bleed manifolds for blown flaps. The final mass-production version was the R-11F2S-300, of 6,200 kg thrust, also made by HAL at Koraput and at Chengdu, China, as the WP-7. The R-13 has a new compressor with higher performance, the R-13-300 ratings being 5,100/6,600 kg.

The RU-19 is a small turbojet used as primary propulsion for RPVs, as a booster for transports and also as a combined booster and APU. Maximum thrust of the RU-19-300 is 900 kg. The R-25 is installationally interchangeable with the R-13 but is completely redesigned, uses much titanium, has a higher pr (about 13) and with afterburner gives 7,500 kg. The R-26 family power MiG-25 versions, this being a steel single-spool engine for Mach 3.2 with a 5-stage compressor (pr about 7), 1-stage turbine and large 3-ring afterburner. Early versions had an uncooled turbine and rating at 10 tonnes, but the production R-31 (R-266) has water-methanol injection in supersonic flight, runs on T-6 fuel (freeze – 62.2°, flash 54.4°) for a rating of 9,300/12,250 kg. The R-31F has afterburning thrust of 14 tonnes.

The R-27 HP spool is derived from the R-25 but has much greater airflow to give ratings of 7 tonnes/10,200 kg. Latest known production engine is the R-29, with 5-stage LP, 6-stage HP, annular vaporizing combustor, single-stage air-cooled HP turbine and 2-stage LP. In most applications it has an outstanding afterburner; typical weight is 1,760 kg, airflow 105 kg/s, pr 12.9, and ratings 8 tonnes dry and from 9,900 to 12,480 kg with afterburner. Tumanskii died in 1973 and his successor has not yet been published.

Turboméca (FRANCE)

Nobody has run a major aero-engine company as long as Josef R. Szydlowski, who with M André Planiol founded Société Turboméca in 1938. Its

business has from the start involved compressors and turbines (originally HS12Y and Z turbochargers), starting in Paris (Billancourt) but moving in June 1940 to Bordes, Pyrenees. The factory was pillaged in 1944, but in 1945 work resumed on a wider front than before, and in 1947 the company designed two gas turbines, the Orédon of 140 hp and the Artouste of 220 hp. These were intended for many tasks including APUs and aircraft propulsion. In 1948 came the Piméné turbojet. These simple yet novel engines established the company as a pioneer of small gas turbines. They had a single-sided centrifugal compressor driven by a 1-stage axial turbine with the blades integral with the disc, with fuel vaporized by being sprayed under centrifugal force from a perforated ring rotating with the main compressor drive shaft. This shaft was of large enough diameter to allow the rotating assembly to be supported between a ball bearing ahead of the compressor and a flexibly mounted roller bearing aft of the turbine. The turbine nozzle vanes (stators) were fabricated from welded sheet, with internal cooling by compressor delivery air. At low throttle settings the fuel delivery was mainly bypassed back to the pump.

These features were repeated in almost all the 55-odd subsequent Turboméca engine types. The only significant changes have been to add first one and subsequently 1, 2 or 3 axial stages upstream of the centrifugal compressor, and use 2-stage turbines with inserted firtree-root blades. The following condensed listing is roughly chronological.

The Orédon was developed to over 160 hp but did not propel aircraft. The Artouste was developed

The Piméné was one of Turboméca's first production engines, and the first small turbojet in the world to be built in quantity.

through the Artouste II (airflow 3.1 kg/s, pr 3.7) of 400 shp at 33,000 rpm to the Artouste III (added axial stage, 4.4 kg/s, pr 5.5) and 3-stage turbine, of 550 hp at 34,000 rpm, and thence to the IIIB flat-rated at 550 hp to 10 km or 45°C. The Piméné weighed 54 kg, had airflow of 2 kg/s, pr 4, and gave 110 kg thrust at 36,000 rpm. The Palas was a scaled-up Piméné (airflow 3.1 kg/s, pr 4) weighing 72 kg and rated at 160 kg at 34,000 rpm. The Marboré I was a further enlargement to 300 kg thrust, flown on the Gemeaux II on 16 June 1951. The same aircraft (redesignated Gemeaux III) flew on 24 August 1951 with the 380-kg Marboré II, and on 2 January 1952 with the production Marboré II (8 kg/s, pr 4) weighing 146 kg and rated at 400 kg at 22,600 rpm. This was the biggest seller, licensed as the J69 to Continental (now Teledyne CAE) of the USA, and also to Israel, Spain, Romania and Yugoslavia. In Britain Blackburn took a licence for all Turboméca engines, effecting considerable redesign but failing to find a market except for the Turmo. This was the first free-turbine engine, derived from the Artouste I in 1950 by adding a single-stage power turbine and gearbox. This was developed into the 400-hp Turmo II (1954) and then to many versions of Turmo III and IV with added axial compressor and 2-stage core and power turbines (typically 5.9 kg/s, pr 5.9), weighing about 300 kg and rated up to 1,610 shp.

The Ossau was a turbojet rated at 800 kg thrust,

Assembling Turboméca Turmo 3C4 engines at Rolls-Royce's Leavesden (Watford) factory in 1972. These descended from one of the first free-turbine engines.

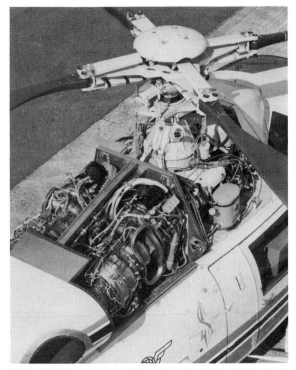

Twin TM333s snugly installed in a Dauphin. This is Turboméca's standard engine in the 1,000-hp class. Like Sikorsky, Turboméca's logo is a winged S, standing for Szydlowski.

with a diagonal (axial-cum-centrifugal) compressor, abandoned in 1952. The Pimédon was an air compressor (among other things, for tip-drive helicopters) derived from the Artouste, used in several versions. The Aspin I, rated at 200 kg thrust, was the first turbofan in the world to fly (Fouga Gemeaux IV, 2 January 1952); it had variable inlet vanes upstream of the single-stage fan, which was driven by a reduction gear from the compressor. The Aspin II, flown five months later, was rated at 350 kg thrust with sfc of 0.52. The Arrius I and II were air compressors larger than the Palouste, and the Autan of 1955 was a Palouste with an added axial stage giving 5.1 pr. The Soulor of 1954 never flew; it was a turbofan of 320 kg thrust. The Gourdon turbojet of 1955 weighed 172 kg and had thrust of 660 kg. The Marcadau turboprop of 400 hp was an Artouste II with 2.332 spur gear. The Gabizo turbojet of 1955 was a sharp upward jump in size, with an added axial stage (pr 5.2, airflow 22 kg/s) to give 1,100 kg thrust at 18,000 rpm or 1,540 kg with Nord afterburner, weights being 265 kg or, with afterburner, 380 kg. The Arbizon turbojet of 1956 had an axial-plus-centrifugal compressor, weighed 104 kg and gave 250

kg thrust at 34,000 rpm. In 1970 the Arbizon IIIB appeared for the Otomat cruise missile, weighing 115 kg and rated at 380 kg at 33,000 rpm, the simpler Arbizon IV weighing 60 kg for 367 kg thrust.

The Bastan of 1957 was the first axial/centrifugal turboprop, weighing 180 kg and initially giving 650 shp plus 78 kg thrust and being developed by 1965 to the 1,048-shp Bastan VII. The smaller Astazou of 1957, like the Bastan a constant-rpm engine with power varied by fuel flow, weighed 110 kg and gave 320 shp plus 40 kg thrust at 40,000 rpm. It became the company's best-selling turboprop, many versions with two axial stages giving powers in the 1,000-hp class, and the Astazou XX (one of many turboshaft versions) having a third axial stage. The Turmastazou, also produced in Double form at 1,775 shp, was an Astazou XIV with an added free turbine. The Astafan of 1969 was an Astazou XIV with 2-stage epicyclic reduction drive to a variable-pitch fan of 6.5 bypass ratio; weighing 230 kg, it gave 712 kg thrust with sfc of only 0.38. The Aubisque of 1961 was a turbofan larger than other Turboméca engines except the Gabizo, weighing 290 kg and with a geared fan, its rating being 742 kg at 33,000 rpm. The

Right *The 1,000th Arriel on view in late 1983. This 700-hp class engine is also made in China.*

Right *Smallest and simplest of the new-generation Turboméca engines, the TM 319 has an advanced centrifugal compressor without added axial stages. The accessories (left) are bigger and heavier than the engine itself.*

Below *Turboméca's standard 2,000-hp engine is the Makila, used in the Super Puma. It is a direct extension of the Turmo but with two extra axial stages.*

Aubisque 6, not made in quantity, was one of the first Turboméca engines to have an air-cooled turbine (the other was the Astazou XVI); thrust rose to 840 kg. The name Orédon was resurrected in 1965 in Mk III form at 350 shp for helicopters, with Rolls-Royce reduction gear; the Orédon IV gave 420 shp at 59,100 rpm.

First run in 1974, the Arriel is one of the leading current turboshaft engines, with 1 axial and 1 centrifugal compressor made in titanium, strong enough for very high rpm giving pr of 9. Weighing 109 kg, the Arriel 1 has a contingency rating of 698 shp and is to be developed in turboprop and 500-kg turbofan versions. Its bigger partner is the Makila, first run in 1977; it has 3 axial plus 1 centrifugal, 2-stage air-cooled gas-generator turbine and 2-stage power turbine, weighs 210 kg and gives 1,875 shp. First run in 1981, the TM333 is a free-turbine engine giving 912 shp, weighing 135 kg, with 2 axial plus 1 centrifugal and 1-stage gas-generator and power turbines. The TM319, first run in 1983, has no axial stage but runs at high rpm for good pr; rating is 443 shp. By 1985 Turboméca had delivered 23,000 engines, and over 14,000 had been delivered by licensees.

Turboméca collaborative engines
(SEE TEXT)

The Adour and RTM.322 are described under RRTI (*qv*). The Larzac is described under SNECMA (*qv*) but is marketed by GRTS (Groupement Turboméca-SNECMA). The TM251 turboshaft for the Agusta A106 was jointly developed with Agusta, who designated the 354-shp engine TAA.230. The MTM385 is being developed by MTU/Turboméca for the Franco-German anti-tank helicopters; rated at around 1,000 shp, it has 2 axial plus 1 centrifugal, 1-stage gas-generator turbine and 2-stage power turbine. MTU and Rolls-Royce are also collaborating on the TM319.

Turbo-Union (INTERNATIONAL)

Turbo-Union Limited was formed in October 1969 to manage the programme for the RB.199 engine.

The MTM 385 is a new helicopter engine in the 1,200-hp class being developed by MTU of Germany and Turboméca of France.

Shares are held in the ratio Fiat 20 per cent, MTU 40 per cent and Rolls-Royce 40 per cent. The RB.199 is a 3-shaft augmented turbofan of unprecedented compactness and light weight; overall length is typically only 127 in (3.23 m) including high-ratio afterburner and integral reverser. All versions have a 3-stage LP (fan), 3-stage IP, 6-stage HP (airflow 155 lb/s, overall pr over 23), annular vaporizing combustor, 1-stage air-cooled HP and IP turbines and 2-stage LP. Typical weight is 1,980 lb (with reverser, 2,390 lb) and ratings are: Mk 101, 8,090 lb dry, 15,950 lb with afterburner; Mk 103 (standard Tornado IDS engine) 9,565/16,920 lb; Mk 104 (lengthened jetpipe and digital control) significantly greater thrust. The Demo 1A engine, run in 1982, leads to further uprating which at high altitude promises to better in-service RB.199s by 40 per cent.

V

Vedeneyev (SOVIET UNION)

Since the early 1960s Ivan M. Vedeneyev (or Vedeneev) has been responsible for modified versions of the Shvetsov ASh-62 (notably the 62M, with a 58-hp spray-pump drive for the An-2M) and Ivchyenko AI-14 (notably the M-14V-26 helicopter engine and the 360-hp M-14P used in Yak and Su aerobatic aircraft).

Volvo Flygmotor (SWEDEN)

Flygmotor was founded in 1930 to make the Bristol Pegasus. Subsequently it made the Twin Wasp R-1830 and DB 605, produced its own 145 hp flat-4 (F-451-A) with bore 125 mm and stroke 105 mm (5.1 litres), and got into gas turbines by producing the DH Goblin and Ghost. Versions of the RR Avon followed, with SFA (Svenska Flygmotor AB) afterburner, as the RM5 and RM6, followed by the RM8, a largely SFA-developed supersonic fighter engine whose basis was the P&W JT8D. Several hundred

Left RB.199s are assembled at each partner. Here MTU fitters complete one of the engines for the first German Tornado, since followed by 1,000 others of later marks. Note the reverser buckets at the top of this remarkably short engine.

RM8As have a 2-stage fan, 4-stage LP and 7-stage HP (145 kg/s, pr 16.5), giving dry/afterburning thrusts of 6,690/11,790 kg for a weight of 2,100 kg (the reverser on the Viggen is part of the aircraft).

These were followed by the RM8B for the JA37 fighter with 3-fan and 3-LP design, revised combustor and new HP turbine, weighing 2,250 kg and with dry/augmented ratings of 7,350/12,750 kg. Volvo took over SFA in 1970, a major current task being a 40 per cent share in the RM12, a Swedish version of the GE F404 to be rated 'in the 18,000 lb class'. Volvo Flygmotor participates in development and production of other GE and Garrett engines.

Checking final details on a Volvo Flygmotor RM8A (rotated upside-down) before despatch from Trollhätten. One is reminded of what Lord Hives said when Whittle told him how simple jet engines were: 'Don't worry, we'll soon design the simplicity out of it!'

W.IV being made in some hundreds at 220-240 hp or 300 hp in racing (high-compression) trim. From 1922 Walter made 17 distinct types of small radial, most with cylinders 105 × 120 mm, with 3, 5, 7 or 9 cylinders and initially with both valves parallel to the cylinder axis; from 1925 these had enclosed valve gear, a feature absent from the Jupiter which the firm made under licence. From 1929 Walter added air-cooled in-line engines, the most famous being the Minor and Mikron. These were inverted engines with steel cylinders, aluminium heads and push-rods to overhead valves; the Minor came with 4 or 6 cylinders of 105 × 115 mm, typical ratings being 105 and 160 hp, and the Mikron had 4 cylinders 90 × 106 mm giving 65 hp. There were 8 and 12-cylinder versions, top of the range being the Sagitta inverted-V-12 of 1937 with 118 × 140 mm cylinders (18.4 litres), weighing 360 kg in direct-drive supercharged form and rated at 500 hp at 2,400 rpm. Today's Avia inverted in-line engines (Minor 6-III, M 137 and M 337) are basically the pre-war Walter designs, and in 1984 the Mikron went back into production at the Aerotechnik works!

W

Walter (CZECHOSLOVAKIA)

A.S. Walter of Prague-Jinonice, was established in 1920 soon after the country gained independence. Its first engines were water-cooled 4 and 6-in-lines, the

This Walter Mikron is typical of the Walters in having all the induction and exhaust piping on the right. A very similar engine is now back in production.

Walter (GERMANY)

Professor Hellmuth Walter, a research chemist in Kiel, experimented with torpedoes driven by concentrated hydrogen peroxide. In 1935 he began designing a rocket engine for aircraft, founded the Hellmuth Walter Kommanditgesellschaft (HWK) and received an RLM (Air Ministry) contract for a 40 kg thrust unit—in modern parlance, a 'bonker'—to be fitted to one wing tip of an aeroplane for roll dynamics research. This led in late 1936 to a unit of 100 kg thrust which boosted an He 72. Next came a profusion of assisted-take-off rockets, and a contract for the R I-203 engine of 400 kg thrust, running on the spontaneous reaction of *T-stoff* (80 per cent peroxide plus a stabilizer) and *Z-stoff* (strong calcium permanganate solution). This powered the He 176 and DFS 194. This was developed into the R II-203 of 750 kg thrust, which powered the Me 163A. In turn this led to the slightly less dangerous R-II-211, with *Z-stoff* replaced by *C-stoff* (30 per cent solution of hydrazine hydrate in methanol). The highly reactive propellants were kept scrupulously separate until, fed by two turbopumps, they met in the reaction chamber. Weight was about 165 kg, and thrust 1,500 kg at sea level rising to 1,700 kg at high altitude. This

entered production as the HWK-509A-2 for the Me 163B. Flight endurance was severely restricted by the propellant consumption of 8 kg/s, so the HWK 509C was developed with a cruise chamber of 300 kg and a main chamber of 1,700 kg. This powered the Ju 248 (Me 263) and Ba 349.

Warner (USA)

Aeronautical Industries Incorporated was formed in Detroit in 1927, changing its name in the same year to Warner Aircraft. Its product, the Scarab 7-cylinder radial, ran in November 1927 and practically swept the board at the 1928 US Nationals. A conventional design, it had cylinders 4.25 in square with enclosed valve gear driven by pushrods behind the cylinders, making for a clean appearance. In 1930 the 90-hp 5-cylinder Scarab Junior was added, followed in 1933 by the Super Scarab 7-cylinder which was made in two sizes, the R-500 (bore enlarged to 4.625, 499 cu in) and R-550 (4.875 in, 555 cu in), with respective take-off powers of 175 and 200 hp. ATA ferry pilots will remember the smoothness of these engines in the Argus (UC-61), but Warner was not included in the massive wartime production. In 1950 the assets were sold to Clinton Machine Corporation.

Westinghouse (USA)

Westinghouse Electric Corporation, one of the world's biggest manufacturers of steam turbines, was

The Westinghouse J30 was the turbojet of 19 in diameter contracted for on the day following Pearl Harbor. It was the first American-designed turbojet to run.

one of the three turbine companies invited to be represented on the Durand Special Committee on Jet Propulsion. By sheer chance Westinghouse decided to recommend a turbojet (Allis Chalmers a turbofan and GE a turboprop). In July 1941 the company received go-ahead on their design, a Navy contract following on 8 December 1941 for the 19A (19 in diameter), an axial engine envisaged as a take-off booster. Under R.P. Kroon the team at South Philadelphia completed the design in 10 months, and had the X19A running on 19 March 1943. This was only the second type of axial turbojet to run outside Germany, and the first American-designed gas turbine to run. It set the pattern in being a simple and robust unit with 6-stage compressor, annular combustor and single-stage turbine. It gave 1,200 lb thrust, and the No 2 prototype was flown as a booster under an FG-1 Corsair in January 1944. By this time designs had been prepared for derived engines to give three sizes, the 9.5 in (diameter), 19 and 24 in. The 9.5A and B were for missiles and aeroplanes, respectively, receiving Navy designation J32. The 19XB for piloted aircraft became the J30, and this had a 10-

Top *The one big success Westinghouse had with turbojets was the J34. This basically simple engine had a nominal (compressor) diameter of 24 in; the J34-WE-34 illustrated was a 3,250 lb engine that powered F2H Banshees.*

Above *In contrast the big Westinghouse J40 caused the greatest upheaval in major aircraft programmes ever resulting from the failure of any engine. This J40-WE-6 was a 7,500 lb engine with no afterburner but still needing a twin-eyelid nozzle. Note the accessories grouped in the V between the aircraft inlet ducts.*

stage compressor and was rated at 1,600 lb. It was tested under a Marauder and powered the FD-1 (later FH-1), 130 of the 261 production engines being made by Pratt & Whitney.

On 1 February 1945 Westinghouse established a separate Aviation Gas Turbine Division. So far this new field had proved a happy experience, and the 24C engine, the largest of the original family of designs, soon found a large and growing market as the J34. This had an 11-stage compressor (originally 50 lb/s, pr 3.65, and in later versions 55 lb/s at 4.35), a combustor of double concentric annular form with 24 downstream burners round the inner flame tube and 36 around the outer, and a 2-stage turbine. Weight was typically 1,220 lb and thrust rose from 3,000 to 3,500 lb at 12,500 rpm. The J34 was conservatively designed and had a long career, a few surviving into the 1980s in P-2 and C-119 booster pods.

Westinghouse maintained a brisk pace of development and in spring 1947 embarked on two new engines, with US Navy BuAer support. These were the J40 (Model 40E, 40 in diameter) and J46, and it was expected that between them they would power almost all Navy fighter and attack aircraft of the 1950s. Because of its great importance the big J40 was timed earlier than the J46, and a prototype XJ40 ran on 28 October 1948. It had a 10-stage compressor, double concentric annular combustor with so-called 'step wall' construction with 16 duplex burners, and 2-stage turbine. In the XA3D bomber the J40 was installed in pods with plain front inlets, with accessories around the underside, but in other applications (XF4D, XF10F, XF3H) all accessories were grouped between the bifurcated inlet ducts ahead of the engine, driven from a pressurized nose gearbox. An early J40 passed its 150-h qualifications test in January 1951, but by this time the J40 had revealed problems of many kinds—aerodynamic, combustion, structural and in detailed mechanical design—which prevented attainment of the design ratings and resulted in extreme unreliability. Development and pre-production engines weighed about 3,100 lb, or 3,620 lb with afterburner, and were planned to run at 7,600 rpm to give a minimum of 7,500 lb dry and 10,500 with afterburner. The best that could be obtained by summer 1952 was 6,000 lb. Douglas managed to switch to the J57 to power the A3D, and later did the same with the F4D. The XF10F finally ground to a halt, leaving the F3H Demon in terrible trouble. This fighter had its fuselage redesigned to accept the more powerful J40-10, but the only engines available were the Dash-8 (unreliable) and the Dash-22 with a constant-speed alternator drive but putting out only 7,200 lb. Flight development was punctuated by accidents and groundings, and the sorry saga finally ended with the last J40 flight on 7 July 1955. Subsequent Demons had the Allison J71 engine.

By 1955 Westinghouse AGT Division was raring to go in its great new facility vacated after the war by Pratt & Whitney at Kansas City. Having been made brutally aware of the difficulties of developing gas turbines, the division had on 15 June 1953 signed a 10-year technical collaboration agreement with Rolls-Royce. Had it simply taken licences, big sales might soon have resulted. Instead by 1955 it was announced that Westinghouse was working on the XJ54 and XJ81 turbojets, respectively in the medium and small thrust classes. The XJ81 was basically the Soar, intended for RPVs and missiles. Rolls were confident that the later Avons could find a large market via Westinghouse, but were horrified to learn that the XJ54 was an Avon scaled down to only 105 lb/s airflow, resulting in an engine weighing 1,500 lb and giving a thrust of 6,200 lb, with low fuel consumption. This was far better than the J46, the only engine of its own design that Westinghouse had left. It was a J34 scaled up to 75 lb/s, with an 11-stage compressor giving pr of 5.2. With afterburner it weighed 2,090 lb, and was rated at up to 6,000 lb. Only a handful were delivered for the F7U-3 and YF2Y, and the XJ54 was never sold at all.

Williams (USA)

Sam Williams is the US answer to Szydlowski, though unlike Turboméca almost all his output has gone into RPVs, targets, reconnaissance drones and cruise missiles. The WR19 2-shaft turbofan, which weighs 141 lb and gives 718 lb (wet rating), was used in Flying Belt, SAVER and WASP manned applications, and the more recent FJ44 was to have powered the Foxjet at a rating of 850 lb. It has since been developed to 1,500 lb, though weighing only 193 lb; an application is awaited.

Wolseley (UNITED KINGDOM)

In 1909 Wolseley Motors, at Adderley Park, Birmingham, began building water-cooled aero engines, starting with a 4-in-line with 4×5 in cylinders, weighing 242 lb and giving 30-36 hp at 1,440 rpm. Almost all subsequent versions were 90° V-8s, with cylinders cast in pairs with screwed-on aluminium jackets, left and right cylinders being staggered to allow two big ends on each crankpin. The 1909 engine had 3.75×5 in cylinders, weighed 350 lb and gave 50 hp at 1,350 rpm. In 1910 metric dimensions resulted in 94×138 mm cylinders, 326 lb and 90 hp at 1,750 rpm, followed by a 126×176 mm engine weighing 635 lb but giving 150 hp at 1,150-1,400 rpm. During the war the firm made the W.4 Viper family of engines derived from the Hispano, adding the 1918 Python to its own design. The market was abandoned until in 1933 a family of simple air-cooled radials was offered, and with sales and service backing by the Nuffield car organization. The A.R.7 and A.R.9 had respectively 7 and 9 cylinders 4.1875×4.75 in, giving 130-145 and 165-180 (later over 200) hp. Few were built.

Wright (USA)

Wilbur and Orville Wright were the first aviators to fly an engine. Their contribution to human flight was, however, mainly in solving the problems of how to build a controllable aeroplane and then learn to fly it. They carried out virtually all the serious research on propellers prior to 1905, and incidentally decided to propel all their early Flyers with two screws turning slowly in opposite directions by bicycle-chain geared drives. They did not want the bother of having to make their own engine, but got such a useless response from established builders that they were forced to do the job themselves. Unfailingly painstaking and methodical, they began by making a 'skeleton engine' with one cylinder, driven off a belt in their bicycle shop in Dayton. After carefully studying its behaviour they decided to go ahead with a complete engine in November 1902.

There is an amazing amount of inaccurate reportage on everything the brothers did. It has been claimed their first engine came from a Pope-Toledo car, that it was all-aluminium, that it had direct fuel injection, that fuel 'was allowed to drip into the intake ports', and that the pilot could speed up or slow down the engine by turning the fuel valve one way or the

The 1903 Wright engine, with crankcase covered in transparent plastic. Left, the chain driving the camshaft and make/break contacts. Right, flywheel and DC generator. In front, the advance/retard controller, four vertical 'valve barrels' linked by the DC feed strap. On top, brass petrol pipe to the air inlet chimney brazed into the cover plate of the hot evaporative chamber.

other, and much, much more. In fact the original engine, in the words of chief mechanic Charlie Taylor, who made almost all of it, 'had no carburettor, no spark plugs, not much of anything. . .but it worked'. It was a 4-in-line lying on its side. The main body was cast in aluminium; the first casting fractured on an early run in February 1903 when the bearings seized. The steel cylinders were 4×4 in (201 cu in), and each screwed-on head had a vertical drum-like section containing a lightly sprung automatic inlet valve above and a strongly sprung exhaust valve below. The latter valves simply let the hot gas escape through surrounding apertures, and were opened by rockers from the chain-driven camshaft which was also geared to the petrol pump. Water jackets surrounded the cylinders, but not the heads, and as there was no forced flow the engine got steadily hotter, losing power. The top of the jacket area was cast in the form of a shallow tray, with a screwed cover, into which was slowly pumped the regular automobile petrol. This chamber was heated before the start, and subsequently got ever hotter, so that it was petrol vapour that mixed with the air induced through a kind of chimney above the cover. The crankshaft was machined from a single slab of steel and ran in 5 white-metal bearings. One end drove the camshaft sprocket and the other the flywheel and two propeller chain sprockets. The flywheel rim provided a friction drive to the low-tension generator feeding simple make-and-break contacts inside the cylinder heads. Weight was almost exactly 152 lb, about 30 lb less than the brothers' target; with all accessories including radiator and piping (but not water) the weight was 174 lb. Power was hoped to be 8 hp, but on its first run on a brake, in February 1903, the engine gave an unexpected 13 hp. Later, at Kitty Hawk, it was tuned to give about 16 hp at a little over 1,000 rpm, but after about a minute or so output always fell to around 12 hp.

For 1904 the brothers designed two 4-in-lines and a V-8, building both the 4-cylinder engines. These had much better water cooling and many other improvements. The installed engine could put out full power for as long as the petrol lasted; indeed it grew more powerful with age: about 15 hp in 1904, 16.64 measured hp in January 1905 and over 20 hp by the end of the busy 1905 flying season! Gradually the brothers refined their engines, arranging the cylinders vertical fashion and adding 6-in-line and V-8 patterns. Cylinder sizes varied, but did not depart far from the 4 in square of the original. Typical 1910 engines were rated at 30 hp (4-cylinder), weighing 170 lb, and 60 hp (V-8). High-tension magneto ignition was fitted, but direct fuel injection was more common than a carburettor, and inlet valves remained automatic. For various reasons the

Above *In 1906 Orville Wright designed the engine which formed the basis of the brothers flying until 1911. To reduce lateral shift of aircraft centre of gravity with and without a passenger the cylinders were vertical. This was the first engine sold to the US Navy for the B-1 seaplane of 1912. Note water pump, magneto and (with crankcase opened) camshaft worm gear to the petrol pump. The only 'production line' of such engines was by Bariquand et Marré in France, with cylinders 112 × 111 mm, rating 30 hp at 1,300 rpm.*

Right *Typical of the last of the original Wright engines was the 6-60 (6 cylinders, 60 hp) of 1913. With stroke lengthened to 4.5 in, this was a total redesign, with long studs holding down water-cooled cast-iron heads and twin carbs feeding on the side opposite to the exhausts. These simple engines were eventually developed to give about 80 hp, weight being about 280 lb.*

1919 to form a new engine company, Wright Aeronautical Corporation, Rentschler becoming president. Wright continued improving the Hispano, the last model being the E-4 Tempest, rated at 200 hp and weighing 480 lb. It also designed two completely new engines. One was the T (Tornado), a large water-cooled V-12 of 1,947 cu in weighing 1,000 lb and intended to replace the Liberty in all heavy US Navy aircraft. Placed under contract in 1921, it began life at 500 hp, went into production in 1922 as the 525-hp T-2, reached 575 hp as the T-3 in 1923, and terminated in the 675-hp T-4 of December 1923. Wright tried to reach 700 hp, and also to produce a racing version to beat the smaller Curtiss D-12, but it was a dead end.

The other new engine was the R-1 nine-cylinder air-cooled radial of 1,454 cu in started in 1920 for the Army. This achieved its design power of 350 hp when run in 1921, but at 884 lb was heavy, and suffered from having poultice heads which had 4 valves and were copied from the Cosmos Jupiter. A famed expatriate Briton, Sam Heron, showed how to make better air-cooled cylinders, and when these were fitted the R-1 became the R-2, later the R-1454, but the Navy handed this to Curtiss in 1923 as lowest-priced bidder. Wright had other problems: the Army settled on the Curtiss D-12, and the Navy announced it would buy no more Wright-Hispanos, and so under intense Navy pressure Rentschler agreed to take over Lawrance in May 1923, at once continuing development of the Lawrance J-1. The J-3 was strengthened

brothers (Orville alone after Wilbur's death from typhoid in 1912) failed to keep abreast of aeroplane and engine technology, and their last engine was made in 1915.

In 1916 the Wright-Martin Aircraft Corporation linked two major US aviation firms to make Hispano-Suiza engines under licence. Large numbers were made of the Wright-Hispano E, a 180-hp V-8 of 718 cu in incorporating improved conrods and detail changes to the cylinder heads and valves. Small numbers were made of the H, of 1,127 cu in giving 300-325 hp. Almost all the Wright-Martins were direct drive. In 1919 Wright-Martin was dispersed, most going to Mack Trucks, but thanks to F.B. Rentschler, a wartime officer concerned with engine production, substantial assets were used in October

and in 1924 the J-4 introduced an improved cylinder in which the hold-down flange was not on the cast aluminium barrel but on the steel liner. At this point the engine was named Whirlwind, and it quickly became the firm's main (almost only) product, still at 787 cu in and giving 215 hp. In winter 1924-25 Wright lost its president, chief engineer and chief designer (see Pratt & Whitney). The depleted team struggled on to improve cooling and fuel economy in the J-4B Whirlwind until, in 1926, E.T. Jones and Sam Heron joined and completely redesigned the cylinders and among other things fitted salt-cooled exhaust valves. The result was the J-5 Whirlwind, which achieved sudden fame when it took Lindbergh to Paris in May 1927.

In mid-1923 the Navy had awarded Wright contracts for two new air-cooled radials. It was increasingly convinced that such engines were preferable, and it had a high opinion of Lawrance, who became a Wright vice-president (and, in 1925, president). First came the P-1, a 9-cylinder engine of 1,652 cu in derived from a Lawrance Army design of 1919 but with a split master rod on a 1-piece crankshaft and with valves fore-and-aft to try to restrict frontal area. In 1924 work began on the P-2, with much better Heron-type cylinders with laterally splayed valves completely enclosed, and with an integral supercharger. The P-1 gave 400 hp, and in early 1925 the P-2, by then named Cyclone, was type-tested at 435 hp. By this time Wright was also working on the intermediate 1,176 cu in Simoon, to give 350 hp, but this was knocked for six by the Wasp; in any case Wright's depleted engineering department did little in 1925.

In mid-1926 the Jones/Heron/Lawrance team, with chief engineer P.B. Taylor, set course again with improved Whirlwinds and Cyclones. The J-6 series of Whirlwinds had cylinders 5 × 5.5 in (bore having previously been 4.5 in), raising displacement from 787 to 973 cu in and power from (typically) 220 to 300 hp at 2,000 rpm. Valve gear and magnetos were moved to the rear, and there were many other changes. These bigger Whirlwinds, known as the R-975 in the new 'displacement' designations, were later joined by the 7-cylinder R-760. They were much used in transports and many other types, being developed to 350 hp (7-cylinder) and 450 hp (9-cylinder), large numbers being made by Continental in World War 2. As for the Cyclone, this became the R-1750 with cylinders 6 × 6.875 in (1,749 cu in), type tested in 1927 at 500 hp, soon raised to 525 hp at 1,900 rpm, and put into production. This began to offer some competition to Pratt & Whitney, and a further shot in the arm was the merger with Curtiss in 1929. The resulting Curtiss-Wright Corporation kept Wright Aeronautical unchanged until 1931, when the engine teams were merged at the Wright plant at

Paterson, New Jersey. Taylor remained chief engineer, but now reported to the new vice-president (engineering) Arthur Nutt, from Curtiss.

In 1932 the united team produced an outstanding new Cyclone, the F. This had bore increased to 6.125 in, giving capacity of 1,823 cu in; it was known as the R-1820. It had better cylinder finning, a forged crankcase and every modern refinement, including provision for a Hamilton variable-pitch propeller, and it was picked by Douglas for the DC-1. Later Cyclones had a near-monopoly in pre-war Douglas Commercials, and it was sheer chance that almost all wartime derived versions had the rival R-1830. In 1930 Wright also embarked on its first 2-row engine, the R-1510 with 14 Whirlwind cylinders (and retaining the name Whirlwind), which slowly developed at 750-775 hp but failed to find a market. An even bigger challenge was the H-2120 for the

Left *When Wright Aeronautical was formed in 1919 this E-4 Tempest was the main product from the Paterson plant. From this point designer George Mead departed from the basic Hispano V-8.*

Below left *One of Mead's fresh designs was the T-4 Tornado, a hefty water-cooled V-12 used in US Navy seaplanes. Its exterior is so clean it looks unfinished.*

Below *Wright's first radial, the R-1 of 1921, was the first successful high-power radial in the United States. Many features came from the British Jupiter, which returned the compliment by later using a close tandem pair of rods to drive the 4 valves in each head.*

Navy. Intended to replace the Curtiss H-1640, this was to be a very slim flat-12 liquid-cooled engine of 1,000 hp which was to lead to the X-4240 of 2,000 hp with four blocks of 6 cylinders. Yet another diversion of effort from 1932 was the Navy V-1800, an advanced V-12 tested at 800 hp but dropped through lack of funds in 1934 (historian Schlaifer notes that this design was sold to the Soviet Union and there taken to 900 hp, but this is news to the Russians!).

Throughout the first half of the 1930s Wright achieved increasing success with the R-1820 Cyclone, which did much to compensate for the failure of new designs. The F-series was made in large numbers at ratings up to 900 hp at 2,350 rpm on 91-grade fuel. By 1937 the G-series featured detail improvements throughout, and increased cooling-fin area to 2,800 sq in per cylinder to give the magic 1,000 hp, and in 1940 the G200 with even deeper fins and twin dynamic counterweights went to 2,500 rpm,

Below In 1924 Lawrance and Mead set Wright on course with the J-4, the first J-series radial to be named Whirlwind. It had 2-valve cylinders, the gear being exposed, and two Scintilla magnetos on the front.

Below right The first production Cyclone was the 525-hp R-1750 of 1927. It had more and deeper fins, enclosed valve gear, and all ignition equipment moved to the rear.

giving 1,200 hp. Vast numbers were made, many of them by Studebaker, one major application being the B-17. By 1944 Wright had begun introducing its W-fin made of aluminium sheet rolled and caulked into fine grooves cut in the barrels. This led to the H-series Cyclone of 1,350 hp at 2,700 rpm, which remained in volume production (not by Wright but by Lycoming and Canadian P&W) for aeroplanes and helicopters throughout the 1950s at 1,525 hp at 2,800 rpm. Last of the single-row engines was the 1942-designed Cyclone 7, marketed from 1945 as the R-1300, based on the Cyclone H but with stroke reduced to 6.3125 in. The 800-hp production engines were licensed to Kaiser-Frazer and then to Lycoming, many going into blimps; the turbocharged version was dropped.

Wright's key to the future was the belated start in November 1935 upon the design of a modern but conventional 2-row engine much bigger than the Cyclone. This was the Cyclone 14, or R-2600, with 14 cylinders of regular 6.125 bore but a new stroke, 6.3125, giving 2,603 cu in. The first run in September 1936 caused Pratt & Whitney to drop its R-2180 and enlarge its own R-2600 into the R-2800, but by this time Wright was about to run a much bigger engine still, the R-3350 Duplex Cyclone, or Cyclone 18. This had 18 cylinders of Cyclone 14 size, giving 3,342 cu in, and while the 14 was aimed at 1,500 hp the 18 was aimed at 2,000. The smaller engine was

certificated at 1,500 hp in June 1937 and was in production by December 1937. PanAm accepted the high fuel price to use 115-grade fuel in the special high-compression version, called Wright 709C-14AC1, fitted to the transatlantic Boeing 314A from late 1938. The R-2600 became a major wartime engine, rated at around 1,700 hp in the B-series, of which over 50,000 were made by Wright at a plant in Cincinnati. By 1944 the BB-series of 1,900 hp went into production at Paterson with a forged steel crankcase and W-finning, but the R-2600 stopped dead at VJ-day.

In contrast, the giant R-3350 grew in importance for almost 20 years from 1938. Design began in January 1936, and the first engine ran in May 1937. It was quickly adopted for all the largest US aircraft, but development was troubled by poor mixture distribution, catastrophic backfires in the capacious induction system, inflight fires and other problems. The crucial application was the B-29, where a very high degree of supercharging was necessary. Each engine had two General Electric turbosuperchargers feeding the gear-driven blower inside the magnesium case on the rear of the steel crankcase. By March 1938 Wright had decided to design its own superchargers (previously assigned to GE) and immense efforts transformed the altitude performance of the R-3350 by 1943. The one thing missing was direct fuel injec-

tion, which would have solved several problems; water injection to boost power on the arduous B-29 take-offs also had to wait until after the war. The wartime R-3350-23 was rated at 2,200 hp at 2,800 rpm and weighed about 2,670 lb. Related engines powered early Constellations, the L-649 and 749 having the 2,500-hp BD series and the A-1 Skyraider the 2,700 hp, 2,900 rpm, CA series (R-3350-26WB), weighing about 2,850 lb. By this time fuel was injected either through holes in the supercharger impeller or direct into the cylinders. Late R-3350s gave 2,800 hp on 115/145 fuel.

During World War 2 Wright was inevitably caught up in the Army programme for unconventional high-power engines. After dropping the H-2120 and V-1800 the merged Curtiss and Wright teams ignored liquid cooling until in about March 1938 work began

Below left *A Second World War R-975-E3 of 450 hp, the last of the Whirlwinds. In most installations the propeller drive was surrounded by an exhaust collector ring. Note that the valve gear has gone to the back.*

Below *The Cyclone 9 family terminated in the mid-1950s with the H-series of 1,525 hp, made possible by the very closely pitched W-fin cooling. Made by Lycoming and P&W Canada for such aircraft as the S2F Tracker, these Cyclones gave just 1,000 hp more than the 1927 Cyclones pictured opposite.*

Left *One of the war-winners was the R-2600, or Cyclone 14. This is one of the BB-series, with W-fin cylinders and up-rated to 1,900 hp for such machines as the TB, and SB2C.*

Right *Chief engineer Wilton G. 'Bill' Lundquist points to one of the 3 turbines, each driven by the white-hot exhaust from 6 cylinders, that turned the R-3350 into the Turbo-Compound. This prototype was on test in January 1949. He never meant it to be the end of the road for Wright Aeronautical.*

for the Army on an 1,800-hp flat-24 to fit inside wings. This was soon found to be an impractical concept, and in early 1939 Wright proposed a compact liquid-cooled radial with 6 rows, each with 7 tiny (51 cu in) cylinders. This was adopted as the R-2160 Tornado, a contract for a 14-cylinder unit being awarded in June 1939. Tests with this showed that the use of multiple high-speed shafts linking the divided crankshaft to the propeller gearbox was sound, and the Lockheed XP-58 and Republic XP-69 were designed to use this promising engine in 1941. But the very short stroke, constriction of inlet manifolds and major problems with valve gear brought termination in 1943, the engineers being desperately needed on the R-3350.

In 1949 president Guy Vaughan retired, the company moved its HQ from Paterson to its vast wartime plant at Wood-Ridge nearby, and under a new president, Roy T. Hurley—reputedly the highest-paid US executive of the period—set course for the future. It realized it needed to get into gas turbines, but chief engineer Bill Lundquist was an experienced and forceful man who was convinced the R-3350 would 'go on for ever'. He was strongly reinforced in this view by the Navy's funding of a significant new development, the R-3350 Turbo-Compound. This harnessed as much as possible (say, 21 per cent) of the energy normally wasted in the exhaust gas, by piping it through 3 blow-down turbines, spaced 120° apart and each fed by 6 cylinders. The turbines had Haynes Stellite blades welded to an Inconel X disc, and via a stainless-steel radial shaft and fluid coupling were geared to the rear of the crankshaft. Each put in about 200 hp at full power, so the engine output rose from some 2,700 to 3,250 hp, without burning any more fuel; weight rose to about 3,600 lb. The Turbo-Compound passed its Navy 150-h test in January 1950, and went into service on R7V Super Connies. Later versions found wide civil and military markets with wet ratings up to 3,700 hp.

These massive assemblages of precision machinery were the pinnacle of the art of the piston aero engine. Lundquist realised too late that 'for ever' was in reality going to end in 1957. Fortunately dynamic Hurley had seen the light in 1950 and bought licences for the Armstrong Siddeley Sapphire and Bristol Olympus. The former was subjected not only to Americanization but also to a redesign process that, for example, replaced the Sapphire's main mid-section diffuser frame, machined from a solid forging, by a dimensionally similar section welded from pieces of nodular (spheroidal-graphite) iron. This all took about two years longer than expected, so that the author, for one, gave up counting the number of engineless F-84Fs parked on the airfield at Farmingdale when he got beyond 100! Delivery of the J65 Sapphire at last got going in 1953, from Wright and from Buick. The Dash-3 version in the F-84F had a Bendix fuel/air starter which screamed up to 100,000 rpm, to crank the J65. It always lit with a

mild explosion, and this was sometimes followed by a much bigger explosion as it disintegrated at over 200,000 rpm, the clutch having failed to disconnect the drive to the accelerating J65! The B-57 engine starter had a solid-fuel cartridge; this poured black smoke as well. A few J65-6 and -18 Navy engines had afterburners.

Director of engineering Jack Charshafian boldly had a vast gearbox designed which, with extra turbine stages, turned the J65 into the T49 turboprop. There was even a TP51A2 commercial version. This began bench testing in December 1952 at 8,000 shp, and at 9,710 ehp began flying in an XB-47D on 26 August 1955, which was about the time that it became evident the T49 was not going to find a buyer. As for the Olympus, Wright tore it apart even more than they had the Sapphire and tried to redesign it into the J67 for the USAF and JT38 Zephyr for the airlines. By the time they were putting the engine together again the J57/JT3 had scooped up the entire market.

Above *Seen here in the more common non-afterburning form, the Wright J65 was a much-redesigned US version of the AS Sapphire. Large numbers were made for B-57s, F-84Fs and A-4s. An experimental J65 was the first engine to fly on liquid hydrogen.*

Left *Perhaps the ultimate classical type of piston aero engine, the Turbo-Compound sought to extract more useful work from each unit of fuel, and incidentally was tightly baffled to make the cooling air work harder as well. Note the pale pipes conveying mixture to the 18 cylinders and the dark pipes taking white-hot exhaust to the 3 turbines.*

The decline of Wright in engines was as rapid as had been that of its partner Curtiss in aircraft. In 1958 it bought a licence for NSU-Wankel RC (rotating-combustion) engines, but efforts to find aviation markets proved disappointing, and the once mighty company sold out in 1985 to Deere (see Lycoming).

Index

Glossary

Afterburner Enlarged jetpipe with variable nozzle in which extra fuel can be burned to boost thrust.
Aft fan Fan, with extra drive turbine, added downstream of core engine.
Aircooled Piston engine whose cylinders have no cooling other than radiant fins.
Annular Pure body of revolution, with no separate flame tubes.
Augmented Boosted by additional fuel injected in bypass air as well as in core gas.
Axial With airflow passing between alternate fixed and moving blades, sensibly parallel to engine axis.

Bank One line of cylinders in V, X, H or W engine.
Bare weight Without accessories, cooling baffles etc.
Big end End of conrod encircling crankpin.
Bipropellant Using two liquids, a fuel and an oxidant.
Block Single casting or a forging containing one bank.
Bore Internal diameter of cylinder.
Broad arrow Three banks in W form.
Bypass Air compressed by fan or LP compressor ducted past core and expelled to atmosphere,

sometimes mixed with core jet in common nozzle.

Can-annular Annular chamber with separate flame tubes.
Capacity Total area of all pistons multiplied by stroke, also called displacement or swept volume.
Centrifugal Accelerating air expelled radially outwards at high speed, diffuser then converting speed to pressure.
Compression ratio Ratio of entrapped volume at TDC (top dead centre) to cylinder volume at BDC (bottom dead centre).
Contraprop Two propellers in tandem rotating in opposite directions about the same axis.
Core Gas producer providing power for turbofan.

Diffuser Expanding duct in which subsonic flow slows down whilst increasing in pressure.
Downdraught Supplied with air from above.
Dry weight Without water, fuel or oil.
Dual ignition With independent sources of spark.
Ducted fan Early term for turbofan.

Emissions Exhaust gas, visible smoke, noise etc.

Fan Multibladed rotor driven by core to produce thrust by accelerating fresh air to rear: if at the front of an engine it also supercharges the core.
Flame tube Thin-walled perforated container(s) inside combustion chamber in which fuel is burned: it controls mixing of dilution (cooling) air.
Flat twin Twin-cylinder horizontally opposed engine (hence flat-4, flat-6).
Free turbine Power turbine driving output shaft only, not connected to gas generator.
Fuel injection Injection of metered supply of fuel into inlet system or (direct injection) of measured doses straight into each cylinder.

Gas generator Basic power-producing part of gas turbine comprising compressor, combustor and turbine.

H engine Cylinder banks form H seen from end: left/right vertically-opposed engines geared to one output.
HP The high-pressure spool in a two-shaft engine forms the physically smaller central portion, with the compressor and turbine separated only by the combustion chamber.

Inlet guide vane One row of radial vanes (blades) immediately upstream of a compressor or turbine.
Inline Engine with single linear bank of cylinders.
Inverted With cylinders hanging down from crankcase.

Jacket Container for cooling water outside cylinder.

Liner Wear-resistant tube inside (usually softer) cylinder, in which piston runs.
Liquid cooled Cooled by liquid other than pure water.
LP The low-pressure spool comprises a compressor upstream of the HP compressor and a turbine downstream of the HP turbine.

Mass flow Measure of airflow through engine.
Monobloc Cast or forged in one piece.
Monocoque With all strength in outer shell.

Overhead valves Normal (poppet) valves in top of cylinder, stems projecting outwards.

PN Performance number (below 100 called octane number), measure of fuel's resistance to detonation (knocking).
Power turbine Providing useful output shaft power

only: not driving compressor.

Pressure ratio Ratio of pressure at compressor delivery to that at inlet.

Propfan Propeller designed for jet speeds, with many broad but thin curved blades like scimitars.

Pulsejet Air-breathing engine with valves or resonant duct giving rapid-fire intermittent combustion.

Ramjet Air-breathing engine which, after being accelerated to high speed by some other means, acts like a turbojet without need for compressor or turbine.

Ratings Permitted maximum powers.

Reverser Device for deflecting engine jet(s) forward, to slow aircraft after landing.

Row Radial cylinders lying in one plane like spokes.

Single-shaft One main rotating assembly; no separate power turbine or LP/HP spools.

sfc Specific fuel consumption is basic measure of efficiency (fuel flow for given power or thrust).

Stage One row of radial blades, all in same plane.

Stator Row of fixed blades upstream of moving compressor or turbine rotor blades (vanes).

Stroke Distance from BDC to TDC (see compression ratio).

Supercharger Blower or compressor increasing density of mixture (in diesel, air) supplied to cylinder.

Turbocharger Supercharger driven by exhaust gas.

Turbofan Core plus large fan driven by extra turbine.

Turbojet Core plus nozzle only.

Turboprop Core plus propeller reduction gear driven by extra turbine or turbine stages.

Turboshaft Core plus output shaft driven by extra turbine or turbine stages.

Two-spool Also called two-shaft, a core having LP and HP spools rotating at different speeds.

Updraught Fed with air from below.

V engine Two banks forming V when seen from end.

Water injection Spray of water, or water/alcohol, to cool airflow and postpone detonation or, in gas turbine, increase density and hence power.

Zero stage Stage added upstream of original compressor.

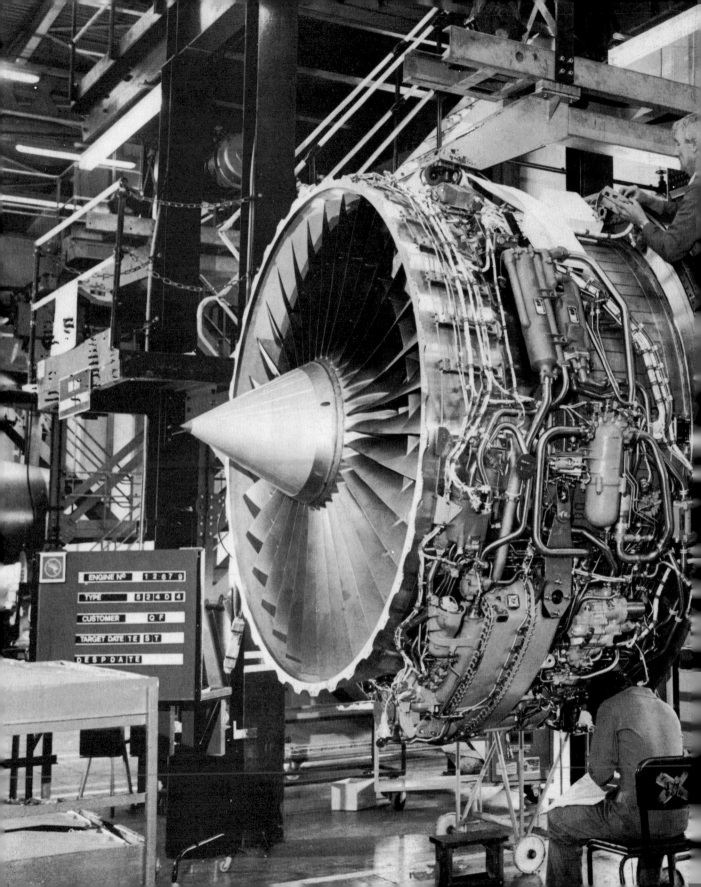